sg · oct · 04

The Frontiers of Europe

The Frontiers of Europe

Edited by

Malcolm Anderson and Eberhard Bort

PINTER

London and Washington

Pinter

A Cassell imprint
Wellington House, 125 Strand, London WC2R 0BB
PO Box 605, Herndon, Virginia 20172

First published 1998
© The editors and contributors 1998

British Library Cataloguing-in-Publication Data
A catalogue record for this book is available from the British Library.
ISBN 1-85567-486-6

Library of Congress Cataloging-in-Publication Data
The frontiers of Europe / edited by Malcolm Anderson and Eberhard Bort.
     p.   cm.
Includes bibliographical references and index.
ISBN 1-85567-486-6
     1. Europe—Politics and government—20th century. 2. Europe—Boundaries. 3. Europe—Foreign relations. 4. Europe—Economic conditions—20th century. I. Anderson, Malcolm. II. Bort, Eberhard, 1954–
D445.F75   1997
327'.094'09049—dc21   97–8653
                                                                                  CIP

Typeset by York House Typographic Ltd, London
Printed and bound in Great Britain by Bookcraft (Bath) Ltd.

Contents

———

Acknowledgements

———

This volume is based on papers presented at a colloquium on the frontiers of Europe jointly sponsored by the Europa Institute and by the International Social Sciences Institute in the University of Edinburgh. The joint conveners were Professor Malcolm Anderson and Professor John Usher. The conference organizer was Eberhard Bort.

For financial support of the colloquium we wish to thank the European Commission, and particularly its representative in Scotland, Mr Kenneth Munro, the Faculty Group of Law and Social Sciences, and the ESRC (through the Research Project on the internal and external frontiers of the European Union – ESRC R000235602).

We are grateful to Anond Lyons of the Geography Department of the University of Edinburgh for drawing the maps in this book.

About the Contributors

Malcolm Anderson is Professor of Politics and Director of the International Social Sciences Institute at the University of Edinburgh. Recent publications include *Frontiers: Territory and State Formation in the Modern World* (Polity, 1996) and, with Monica den Boer, Peter Cullen, William C. Gilmore, Charles D. Raab and Neil Walker, *Policing the European Union* (Clarendon, 1995).

Andrew Barker is Head of the German Department at the University of Edinburgh. His latest publication is *Telegrams from the Soul: Peter Altenberg and the Culture of Vienna* (Camden House, 1996).

Didier Bigo is Maître de Conférences at CERI (Centre for International Studies and Research), at Fondation Nationale de Sciences Politiques. He has recently published *Polices en réseaux* (Presses des Sciences Po, 1995) and 'The European Security Field: Stakes and Rivalries in a Newly Developing Area of Police Intervention', in M. Anderson and M. den Boer (eds) *Policing Across National Boundaries* (Pinter, 1994).

Eberhard Bort is a Research Fellow at the International Social Sciences Institute at the University of Edinburgh. Together with Malcolm Anderson, he edited *Boundaries and Identities: The Eastern Frontier of the European Union* (ISSI, 1996). He is also the editor of two books on Irish drama: *Standing in their shifts itself . . . : Irish Drama From Farquhar to Friel* (European Society for Irish Studies, 1993) and *The State of Play: Irish Theatre in the 'Nineties* (Wissenschaftlicher Verlag Trier, 1996).

John Breuilly is Professor of Modern History at the University of Birmingham. His publications include *The State of Germany: The National Idea in the Making, Unmaking and Remaking of a Nation-State* (Longman, 1992), *Nationalism and the State* (Manchester University Press, 1993) and *The Formation of the First German Nation-State, 1800–1871* (Macmillan, 1996).

Anthony P. Cohen is Head of Department of Social Anthropology at the University of Edinburgh. He edited *Symbolising Boundaries: Identity and Diversity in British Cultures* (Manchester University Press, 1986) and published *Self Consciousness: an Alternative Anthropology of Identity* (Routledge, 1994).

Peter Cullen holds a Jean Monnet chair in the Europa Institute at the University of Edinburgh. Together with Klaus Goetz he contributed to and edited *Constitutional Policy in Unified Germany* (Frank Cass, 1995).

Michel Foucher is Director of L'Observatoire Européen de Géopolitique at the University of Lyon II. His publications include *Fronts et Frontières, un tour du monde géopolitique* (Fayard, 1991, Second edition); he edited *Les Défis de sécurité en Europe médiane* (FED/Documentation Française, 1996) and, together with I. Oyarzabal, *Visions of Europe* (Foundation BBV, 1996).

Russell King is Professor of Geography at the School of European Studies of the University of Sussex at Brighton. He edited *The New Geography of European Migrations* (Belhaven Press, 1993) and, together with L. Proudfoot and B. J. Smith, *The Mediterranean: Environment and Society* (Arnold, 1997).

Joseph A. McMahon is Reader in Law at The Queen's University of Belfast. His major publications include *Agricultural Trade, Protectionism and the Problems of Development: a Legal Perspective* (University of Leicester Press, 1992) and *Education and Culture in European Community Law* (Athlone Press, 1995).

Peter-Christian Müller-Graff is Director of the Institute of German and European Social and Economic Law at the University of Heidelberg. He edited *East Central European States and the European Communities: Legal Adaptation to the Market Economy* (Nomos, 1993) and, more recently, together with Ferenc Mádl, *Hungary – From Europe Agreement to a Member Status in the European Union* (Nomos, 1996).

Tom Nairn is a Honorary Fellow at the Department of Sociology at Edinburgh University. His publications include *The Break-up of Britain: Crisis and Neo-nationalism* (Verso/NLB, 1977) and *The Enchanted Glass: Britain and its Monarchy* (Radius, 1988).

Raimondo Strassoldo is Professor of Sociology at the University of Udine. His books include *From Barrier to Junction: Towards a Sociological Theory of Borders* (ISIG, 1970); together with G. Delli Zotti he edited *Co-operation and*

Conflict in Border Areas (Angeli, 1981). His most recent publication is 'Ethnic Regionalism vs the State: The Case of Italy's Northern Leagues', in L. O'Dowd and T. M. Wilson (eds) *Borders, Nations, States* (Avebury, 1996).

John A. Usher is Director of the Europa Institute at the University of Edinburgh. Among his publications is *The Law of Money and Financial Services in the EC* (Oxford University Press, 1994).

Neil Walker is Professor of Law at the University of Aberdeen. He published 'European Integration and European Policing', in M. Anderson and M. den Boer (eds) *Policing Across National Boundaries* (Pinter, 1994), 'European Policing in Transition', in O. Marenin (ed.) *Policing Change, Changing Police: International Perspectives* (Garland, 1996) and 'Co-operation in Justice and Home Affairs: the Politics of the Third Pillar', in R. Bideleux and J. Bradbury (eds) *The Development of the European Union: Policy Integration in the 1990s* (Macmillan, 1996).

Illustrations

Figures

Tables

Chapter 1

European Frontiers at the End of the Twentieth Century: An Introduction

MALCOLM ANDERSON

The general context

European frontiers became a centrally important subject after the unexpected dismantling in 1989 of the Berlin Wall. Frontiers had been removed from the European political agenda, as several contributors to this volume point out, by the Cold War and the long domination of the continent of Europe by two hegemonic powers, the USA and the USSR. European frontiers were therefore stable between the late 1940s and the late 1980s – until the political cataclysm of 1989 and its aftermath resulted in, as Michel Foucher emphasizes in the final chapter, the drawing of 8,000 miles of new international boundaries in Central and Eastern Europe.

In the long period of territorial stability, questions of frontiers and boundaries were, of course, raised in a variety of forms but these were usually regarded as marginal to the great questions of international politics. There were internal identity/minority-related boundary-questions in Europe (Northern Irish, South Tyrolese, Basques, Slovenes, Corsicans) but separatist and even autonomist parties were widely dismissed as pursuing impractical chimera. More down-to-earth initiatives were in the area of cross-border co-operation between local and regional authorities, which developed particularly in the 1970s, inspired by infrastructure planning, and anti-pollution and environmental campaigns. Transfrontier labour markets emerged, sometimes at the frontiers of two EC Member States as on the Rhine frontier, sometimes between a Member State and a neighbouring state such as at the Basel and Geneva frontiers. Transfrontier flows of people, goods and information increased to levels which raised vague concerns about the ability of states to control and police their frontiers and the activities which took place on their territories.

From the middle of the 1980s, change accelerated in ways which stimulated the emergence of new perspectives on frontiers and frontier policy. The dismantling of frontier controls was implicit in the four freedoms contained in the Treaty of Rome – freedom of movement of goods, services, capital and labour. But progress was slow until the 1985 Single European Act and the European Commission's White Paper, which followed it, containing a long list of measures to achieve a completely integrated market by the beginning of 1993. Some countries forged ahead to plan for the consequences of the dismantling of frontier controls by signing the Schengen Agreement of 1985 (followed by the Implementation Agreement of 1990). All countries of the EU eventually signed up to these agreements with the exception of Britain and Ireland. Increased co-operation in Justice and Home Affairs was an important element of the 1992 Treaty of European Union (popularly known as the Maastricht Treaty) as a 'compensation' for the loss of security which frontier controls were alleged to provide. On another wavelength, border regions became target zones for EC regional policies through the Regional Fund/Cohesion Fund and, more specifically, the two Interreg programmes which funded transfrontier initiatives at the local level. These were directed towards reducing the long-standing psychological and material effects of proximity to international frontiers.

Changes within the EC were in train from the middle 1980s but they developed within the context of a transformed external environment. Germany was reunited; 'alienated borderlands' on the eastern frontier of the EC/EU were transformed into 'open borders'; new members were accepted in 1995 by the EU (the neutral countries Sweden, Finland and Austria) and further extensions to the east were envisaged; direct military threat to western Europe was removed and the whole framework of collective security in Europe had to be reconsidered. These changes brought new opportunities and problems. In the immediate aftermath of the dismantling of the Berlin Wall an optimistic euphoria prevailed in broad sections of western European public opinion but this gave way to anxieties, partly government led, about actual and potential problems. These included alarm at the possible movement of people escaping poverty and disorder in the East (as well as from the southern shores of the Mediterranean), unfair competition from low labour cost countries in agricultural and industrial goods, cross-border crime and new routes into Europe for illegal drugs, and the spread of effects of political disorder either through the influx of refugees or through foreign terrorist groups operating within the EU.

The general background to these anxieties was the feeling of navigating through uncharted waters. The geopolitical certainties of the Cold War were replaced by a whole series of uncertainties. Were the 8,000 miles of new frontiers stable? Would a resurgence of nationalism have uncontrollable consequences and result in further modification of frontiers? Could some of the newly independent states survive? What sort of security guarantees could they and should they be given? What is the role of the United States in the new security situation? Did membership of the EU imply a security guarantee? Is it desirable to pursue a policy of maintaining a buffer state system between the EU and potentially unstable regions further to the east? How far east should the EU expand? This raised the old, unanswerable questions – what is Europe? Where is Europe? Is there a core Europe (perhaps composed of the original signatories to the Treaty of Rome) surrounded by a series of peripheral Europes? Is Turkey genuinely a potential member of the EU or are there deep cultural divides marking some frontiers between Islam and Christendom and even between Orthodox Christendom and Western Christianity? Did the huge economic gulf between the EU and its neighbours to the east and the south exacerbate these cultural differences and make for an inevitable increase in tension and violence?

The politics of cultural anxiety seemed once again to feature large in European politics with mysterious and alien threats against the values of West European societies – illegal immigrants, mafias, terrorists, Islamic fundamentalists – and looming behind these the general threat of globalization destroying European cultural values. The first category of threats involved a redefinition of European security and they could, many argued, be kept under control by increased co-operation between police and security forces and improved intelligence and surveillance techniques. The second was more difficult because inexorable pressures of integrated world markets, new technologies and cultural homogenization could not be controlled. Globalization seemed to imply that all frontiers, not only those of Europe, would be effaced. Europe could not opt out of the global economy, but frontiers as lines of cultural defence had increased salience, and were regarded in some countries, notably France, as instruments behind which some kind of protection against external economic forces could be mounted. There is a large element of myth-making behind these anxieties and proposed solutions; none of the contributors to this volume share the outlook implicit in them. But changes in the global context have raised issues about the security and the protective functions of frontiers which will not quickly be resolved.

The political science of frontiers

The development of new types of frontiers within the EU and new external frontiers poses serious problems for the contemporary analysis.[1] The general role of frontiers in contemporary political life has seldom been explicitly analysed by political scientists. This is partly because boundary effects on the behaviour and values of the populations enclosed by them are almost impossible to assess, let alone measure. Indeed, attempts to measure them seem shallow and produce obvious results which derive directly from the assumptions on which they are based.[2] More significantly, there are basic differences of view about frontiers in the historical and political science literature which are seldom made explicit.

Some historians and political scientists regard the characteristics and functions of frontiers as dependent on the internal organization of societies, and the way in which political power is exercised in the core regions of states. Debates between realist, pluralist, Marxist and interdependence theorists in international relations arise out of different views about the nature of the state; frontiers are regarded as epiphenomena whose role and function are dependent on the core characteristics of the state. For others (including most political geographers), the characteristics of the frontier are fundamental influences on the way a society develops and on the political options open to it. The vast literature specifically on frontiers – the historical classics of Turner and Prescott Webb, geographical treatises, borderland studies and a very diverse literature on border disputes – gives little guidance for the examination of the great changes now occurring as states are easing frontier controls or finding that they cannot use frontier controls to police and control their territory.

Definition of the international frontier

Frontiers can be analysed (and, in normative political theory, criticized) in the same way as other political institutions and processes. Frontiers are not simply lines on maps, the unproblematic givens of political life, where one jurisdiction ends and another begins. As institutions they are established by political decisions and regulated by legal texts. The frontier is the *basic* political institution: no rule-bound economic, social or political life in advanced societies could be organized without them. This primordial character of frontiers is embodied in public international law by the 1978 Vienna Convention on State Succession. When a state collapses, the agreements concerning its frontiers remain in force – frontiers are therefore regarded as prior to the reconstitution of a state and are recognized to be a prerequisite of this reconstitution. Also, frontiers define, in a legal sense, the identity of individuals because the conditions for claims to nationality and exercise of rights of citizenship are delimited by it.

Within its frontiers, the state is a sovereign jurisdiction and the Weberian doctrine of the monopoly of the legitimate use of force on its territory is still almost universally recognized. The doctrine of sovereignty remains a central part of thinking about states and relations between them. The doctrine implies that states have absolute control over their territories and can impose this control at their frontiers. The claim of the modern State to be 'the sole, exclusive fount of all powers and prerogatives of rule'[3] could only be realized if its frontiers were made impermeable to unwanted external influences. But this view of the frontier of the sovereign state is not part of an immutable natural order. Different conceptions of the frontier as an institution existed before the modern sovereign state and other kinds will emerge after its demise. In this regard, there are now signs of seismic political change.

Frontiers are part of political processes with four defining dimensions. First, frontiers are instruments of state policy because governments have attempted to change, to their own advantage, the location and/or the functions of frontiers. Although there is no simple relationship between frontiers and inequalities of wealth and power, government policy on frontiers is intended to promote interests of populations or groups protected by the frontiers. Second, the policies and practices of the state are constrained by the degree of de facto control which the government exercises over the state frontier. The incapacity of governments in the contemporary world to control much of the traffic of persons, goods and information across their frontiers is changing the nature of both States and frontiers. Third, frontiers are basic markers of identity – in the twentieth century usually of national identity, but political identities may be larger or smaller than the nation-state. Frontiers, in this sense, are part of political beliefs and myths about the unity of the people, and sometimes myths about the 'natural' unity of the territory. These 'imagined communities', to use Benedict Anderson's phrase concerning nations,[4] are now a universal phenomenon and often have profound historical roots. They are linked to the most powerful form of ideological bonding in the modern world – nationalism. Imagined communities may transcend the confines of the state, and myths of regional, continental and hemispheric unity have also marked boundaries between friend and foe.[5] But myths of unity can be created or transformed with remarkable rapidity during wars, revolutions and political upheavals.

Lastly, the 'frontier' is a term of discourse. Meanings are given both to frontiers in general and to particular frontiers, and these meanings change from time to time. 'Frontier' is a term of discourse in law, diplomacy and politics, and its meaning varies depending on the context in which it is used. In the scholarly languages of anthropology, economics, history, political science, public international law and sociology, it has different meanings according to the theoretical approach used. Sometimes scholarship is the servant of political power when frontiers are in dispute.[6] For people who live in frontier regions, the term 'frontier' is associated with the (often irksome) rules and restrictions imposed by frontiers, and is also suffused with popular symbolism based on how the frontier is perceived, as either 'barrier' or 'junction'.[7] The layers of discourse – political, scholarly, popular – always overlap but never coincide – divergent mental images of frontiers are an integral part of frontiers as processes.

What frontiers are, and what they represent, is constantly reconstituted by human beings who are regulated, influenced and limited by them. But these reconstructions are influenced by political change and the, often unpredictable, outcome of great conflicts, against a background of technological change. The military technology developed in the closing stages of the Second World War altered the strategic significance of control of territory, the military independence of many sovereign states was drastically reduced and the frontiers of these states became indefensible. Now, with instantaneous communication of information, 'information sovereignty' has been lost; the development of mass, rapid and inexpensive systems of transport have resulted in all developed countries in frontier crossings by individuals annually numbering more than the total population of the country concerned.

A new research agenda

The linear frontier, delimiting sovereign authority and within the confines of which governments attempted to monopolize control of all legitimate use of coercive power, was a European invention. It spread through the rest of the world through imperial domination or influence of the European powers. A renewed focus on European frontiers – in particular the frontiers both internal and external of the European Union – is justified by evidence that changes to the characteristics and functions of these frontiers has been more radical than elsewhere. There may indeed be implications emerging from these changes for other regions of the world.

Currently, there are five main headings under which arguments may be grouped relating to the internal and external frontiers of the European Union.

1. The administration and policing of frontiers.
2. The attitudes towards and perceptions of frontiers, particularly as instruments of 'cultural defence'.
3. The development of institutions and practices of transfrontier co-operation.
4. Conflicts of interest over frontiers or created by frontiers.
5. The exploitation of frontier or territorial anomalies.

Much political analysis conducted has approached topics under these headings in a piecemeal manner. The time has now come for a more general approach and the chapters in this volume make a significant contribution to this.

As I have written in another place – 'Frontiers are inseparable from the entities they enclose.'[8] The degree of administrative, regulatory, political, social and cultural integration which has been achieved in the European

Union can be assessed by examining changes in boundary functions. These changes are not even over the whole territory of the European Union, and indeed some of the external frontiers of the EU may be more 'open' in some senses than the frontiers which divide Member States. Global economic and technological change probably affects some European frontiers more than others but assessments of this, at the moment, can only be speculative.

A number of centrally important empirical questions remain, on which we lack systematic information. Amongst these are:

- Is there a change in the intensity and type of exchanges across EU internal and external frontiers?
- To what extent do internal frontiers continue to act as barriers to political and social exchanges?
- Have important breaches been made in the territorial principle of exclusive control by the Member States of the European Union?
- Does the external frontier have less the character of a linear boundary and more of a broad zone (*limes* or the classic conception of the imperial frontier) where the influence of the EU (and Member States) gradually fade with distance from the frontier?
- Do frontiers impose costs (pollution, provision of additional services, policing and management of the frontier, etc.) and are these costs consistently dissimilar for the external and internal frontiers?

This volume sets out to provide some answers and ask more questions about the agenda set out above. Some progress has already been made on the agenda by the contributors to this volume and published elsewhere.[9] The contributors come from France, Germany and Italy, as well as the UK, which provides an international flavour and a diversity of intellectual perspectives. This is important because the perceived importance of frontiers and frontier functions is strongly influenced by location. Even scholars are influenced by the place in which they live and work whether it is Paris, Belfast, Heidelberg, Lyon, Friuli or Edinburgh. The preponderance of Scottish contributors (particularly among the lawyers) is an indication of close associations between Scottish and continental European legal and social science traditions. Peripheral locations also sometimes encourage an interest in frontiers.

The contribution by Peter-Christian Müller-Graff which follows these introductory remarks is an essential review of legal definitions of frontiers and responsibility for frontiers in contemporary Europe. In a sharply contrasting analysis Anthony Cohen, as a social anthropologist concerned more with perceptions and behaviour than with institutions, develops the important theme of how and why cultural identities become politicized. He also alludes (as does Michel Foucher) to the confusing matter of linguistic usage – the words 'frontier', 'border' and 'boundary' are used in different, indeed contradictory, ways in different disciplines; the rigorous conventions of geographers are not respected in, and are perhaps not appropriate to, other disciplines.

We do not forget that 'the owl of Minerva flies at dusk' and that historical processes cannot in some sense be understood until we know the end of the story. However, for a satisfactory provisional understanding a historical perspective is essential. In this collection the historical evolution of mentalities and identities in German-speaking Europe, and their relationship with frontiers, is presented by John Breuilly who is particularly concerned with the development of the notion of citizenship; and by Andrew Barker, in a more narrowly focused way, from the point of view of the cultural historian, on Austrian identity.

The difficult eastern and southern frontiers of the European Union are the subject matter of the next three chapters. Raimondo Strassoldo, a sociologist with a long and atypical interest in borders and borderlands studies, gives a rich synoptic view of the problems of the borderlands of the Italian north-east as well as the various ways in which they have been approached. This frontier has a complex and violent history and is the point of contact between the three great language groups of Europe – the Latin, the Slav and the Germanic. Eberhard Bort presents a study of the more northern section of this difficult frontier which has undergone such dramatic political change since 1989, to suggest that functional change of frontiers introduces a series of new items on to the political agenda. Russell King, a geographer, adopts a general social science approach to the problematic southern façade of the European Union; the Mediterranean, the 'centre of the earth', has often been a unifying sea but now seems to represent the great frontier between the developed 'north' and the overpopulated and poor 'south'.

Tom Nairn examines an often neglected feature of the international system in which frontier functions are changing and capital flows cannot be controlled – the increasing prosperity and importance of micro-states. Didier Bigo approaches the topics of frontier controls, migration and security from the intellectual background of late twentieth-century French social science; he analyses the representations of the world implicit in the discourses of key groups in the public debate on these subjects. Neil Walker, although a Professor of Public Law, also adopts a general social science approach to the complex set of problems of interpretation of the development of transfrontier police co-operation and institutions in Europe, and what these mean in terms of transformation of the state system and for the doctrine of territorial sovereignty.

Legal analysis is strongly represented by three contributors: Peter Cullen on frontier issues before the European Court of Justice (ECJ), Joseph McMahon on the protection of culture and John Usher on language. Peter Cullen shows the various, but up to the present limited, ways in which frontier issues have come before the ECJ – but the decisions of the Court can clearly have very considerable influence on the conduct of business and on perceptions of frontiers within the EU. Defence of cultural diversity is often regarded as a cloudy, well-meaning slogan, but Joseph McMahon shows that, if it is to have practical effect, legal instruments and a precise idea of what we are talking about are essential – this kind of approach should have greater influence in the social sciences than it has at the moment. Language frontiers are the most obvious barriers to communication in contemporary Europe and language difficulties are evident in interpretations of the most down-to-earth of policy decisions, as John Usher skilfully illustrates.

Finally, a distinguished French specialist in geopolitics provides a coda for the whole volume touching on concepts of the nation-state, issues of state sovereignty, the idea of *limes* and frontier strategy, the potential erosion of borders, integrated borderlands (Euregios and other devices now in process of being translated eastward), border controls in Europe challenged by the global trends towards interdependency, integration of markets and new communications technologies. He also echoes Didier Bigo in pointing to the delocalization of the frontier – crime-fighting beyond frontiers at place of, or closer, to origin of crime (or illegal migration). The broad sweep Michel Foucher presents will have to be revisited again and again in the coming years.

Notes

1. For a more extensive treatment see M. Anderson (1996) *Frontiers: Territory and State Formation in the Modern World*, Cambridge: Polity Press.
2. For example, J. R. Mackay (1969) 'The Interactive Hypothesis and Boundaries in Canada: a Preliminary Study', in B. J. L. Berry and D. F. Marble (eds) *Spatial Analysis: a Reader in Statistical Geography*, Englewood Cliffs: Prentice Hall, pp. 122–9. But when statistical analyses of boundary effects are firmly embedded in historical accounts of the development of boundary relations, a much enriched account may result. See Z. Rykiel (1985) 'Regional Integration and Boundary Effect in the Katowice Region', in J. B. Goddard and Z. Taylor (eds) *Proceedings of the 7th British–Polish Geographical Seminar*, 23–30 May 1983, Warsaw, pp. 323–32.
3. G. Poggi (1978) *The Development of the Modern State*, London: Hutchinson, p. 92.
4. B. Anderson (Revised edition, 1991) *Imagined Communities. An Enquiry into the Origins and Spread of Nationalism*, New York: Verso.
5. W. Connor (1969) 'Myths of Hemispheric, Continental, Regional, and State Unity', *Political Science Quarterly*, **84**(4), pp. 555–82.

6. The relationship between the Nazi regime and an academic discipline (*Ost-forschung* – research on the East) is a chilling example. See M. Burleigh (1988) *Germany Turns Eastwards. A Study of Ostforschung in the Third Reich*, Cambridge: Cambridge University Press.

7. R. Strassoldo (1970) *From Barrier to Junction: Towards a Sociological Theory of Borders*, Gorizia: ISIG.

8. Anderson, *Frontiers*, p. 178.

9. For example, M. Anderson *et al.* (1995) *Policing the European Union. Theory, Law and Practice*, Oxford: Clarendon Press, especially Chapter 6 'Frontiers and Policing'; D. Bigo (1995) *Polices en réseaux*, Paris: Presses des Sciences Po; M. Foucher (Revised edition, 1991) *Fronts et frontières*, Paris: Fayard; R. King (ed.) (1993) *Mass Migration in Europe*, London: Belhaven Press; R. King, L. Proudfoot and B. J. Smith (eds) (1997) *The Mediterranean: Environment and Society*, London: Arnold; P.-C. Müller-Graff (1993) *East Central European States and the European Communities: Legal Adaptation to the Market Economy*, Baden-Baden: Nomos.

Chapter 2

Whose Responsibility are Frontiers?

———

PETER-CHRISTIAN MÜLLER-GRAFF

Introduction

'Whose responsibility are frontiers?' At first sight this question may come as a surprise in an international context. Usually the notion of frontiers is more connected with the idea of competences to exert power and to restrict access from other countries than with obligations and responsibilities. Hence, even asking the question shows a remarkable change in approaching the phenomenon of frontiers.

Whose responsibility are frontiers? In order to give a tentative answer to this question the concept of 'frontier' must be analysed, then the meaning of 'responsibility for frontiers' considered, and the question of who is responsible can then be addressed.

The concept of frontiers

As far as the concept of 'frontier' is concerned, at least three levels of perception should be taken into consideration: the common sense perception of the phenomenon; the concept in public international law; and the meaning in European law.

The common sense perception of frontiers

Common sense perceptions of frontiers have different features and different contexts.

The variety of terrestrial borders

Frontiers may take the form of a terrestrial borderline delimiting one State, like a landed estate, and separating it from surrounding territory which does not belong to that State. Such a border can have different appearances and features. It may have the character of an almost insurmountable obstacle to

everyone who wants to leave or to enter a State. Such an obstacle might take
the form of geographical features like deserts and high mountains such as
the north-western frontiers of China. It can also be created by artificial
means such as walls, barbed-wired fences, watchtowers, mines, shooting
devices and manpower as was the case before 1990 of land-locked Czechoslovakia in relation to West European states. At the other extreme, frontiers
can fade into abstract lines that stand out neither in the landscape as, for
example, between the Netherlands, Belgium and Germany, nor in sharp
cultural separation as, for example, in rural areas along the state borders
between Germany, Luxembourg and France.

The natural limits of islands

A different appearance and experience of frontiers occurs in States or
territories that are islands. Here the often rather artificial, relative and
ambivalent character of a terrestrial borderline is absent and probably little
understood by islanders: a line drawn in and through the landscape,
separating states and their powers, but also regularly cutting through pre-
existing links of landscape, economic region, culture, language, peoples and
family relations; this line is not made for eternity when considering continental European history where a change of frontiers in the last 200 years
on a map looks like a paper pattern for a lady's dress in fashion magazines.

This experience and perceptions of it differ from frontiers of an island-state
or a coastline region which are both clearer and more obscure at the same
time. Wherever the cliffs and the beaches of the island touch the water, there
is a perception of borders as framing a natural habitat as well as providing a
natural protection against other human beings. But maritime borders are
difficult to visualize in their new manifestation as economic exploitation
zones granted by international law. The absence of neighbouring states with
shared land borders might explain occasional difficulties which islanders
have in understanding the genuine challenges of coping with terrestrial
borders and frontiers in a peaceful and mutually beneficial way.

The artificial frontiers of boundless skies

Again, different perceptions and experiences of frontiers concern the skies.
Although it is impossible to demarcate boundaries, the skies have been
partitioned by artificial frontiers through the notion of airspace. The emergence of air transport has also created airport-frontiers inside a state in order
to ensure that the physical presence of a person or a good is not equivalent to
its legal presence in the State. This observation leads to the question of how
a frontier is perceived in international public law.

The concept of frontiers in international public law

Concept

In public international law, frontiers appear as part of the three classic essentials of statehood: people, territory, and authority. Although not expressly listed, frontiers form the necessary delimiting element of territory and of sovereign authority. Hence in international public law frontiers are thought of not so much as separating territories but as physically confining a State, not unlike the American idea of the 'last frontier' of settlement.

Consequences

This concept has several consequences. It raises the question of who is responsible for the definition of frontiers. Questions concerning the concept of sovereign authority and of rights and obligations are also posed.

The responsibility for the definition of frontiers, as confining State authority *in principle*, rests with the state, as long as there is no neighbouring state. However, neighbouring states usually exist, and it is in principle the sole responsibility of neighbouring states to decide on the line between them. Their agreement is deemed to be valid *erga omnes*, demanding full respect by all other nations. Most borders have been fixed by neighbouring states, mutually and autonomously agreeing on the limits of their territorial authority.

However, boundaries have also been defined by a *larger number* of states with an interest in the delimitation of territorial authorities in a region. This multilateral process of negotiating borders and establishing a new territorial order has been used in Europe in the aftermath of great conflicts – for example after the Thirty Years War, when the 1648 Peace of Westphalia created a new territorial order, or after the Napoleonic Wars, when the 1815 Congress of Vienna settled territorial disputes, or after the First World War, when the 1919 Treaty of Versailles tried, but did not succeed, in establishing a stable international order. When boundaries are defined by multi-party treaties, frequently several or all contracting parties act as guarantors of a frontier and are therefore responsible for seeing that it is respected.

The last consequence of the concept of borders as the geographical limits of state authority concerns customary international law delimitation of frontiers where agreements do not exist. For these cases, customary international law provides a series of rules to identify the satisfactory borderline, and those rules are rooted in the primacy of state authority. Traditional respect by states for a certain line may qualify as a source of law. Also, tacit acquiescence of a neighbour to a long-standing practice will be presumed to constitute acceptance of a status quo. Similarly, the principle of effective control over a contested territory might be taken as a right to determine a border. If none of these indicators are available, rules relating to geographic characteristics such as the *Thalweg* of navigable rivers may be the basis of a border.

The core characteristic of sovereignty is traditionally related not to responsibility but to power. As a legal principle, sovereignty describes the monopoly of a state of the legitimate use of force within its territory. This includes the power to set rules by which people are allowed to enter the territory which necessarily includes the right to establish border controls. But the power to set rules of exit which determine whether people are allowed to leave the country is disputed for well-founded reasons.

The concept of frontiers as delimiting state authority has consequences for rights and obligations arising in the context of frontiers.

In principle, a sovereign state has the competence to regulate inward and outward movement; hence a state has the power to refuse the import of goods or the entry of persons to its territory. According to the traditional concept of sovereignty, states are even empowered to close their borders completely to imports and entry. However, there are limitations to this power to regulate inward and outward movements.

As far as persons are concerned, the main limitation to closing borders derives from the International Covenant on Civil and Political Rights which stipulates the freedom to leave a country. On the other hand the right of refugees to enter a country of their choice is a contentious issue and partially subject to the Geneva Convention on Refugees.

In relation to the transboundary movement of goods, there is no customary right to transport goods across borders. Border controls on imports aim to ensure the payment of duties and taxes and compliance with the national standards on health and safety. However, in the context of the European Community the objective of the internal market implies an area without internal frontiers in which the free movement of goods, persons, services and capital is ensured in accordance with the provisions of the EC Treaty.[1] The aim of the White Paper of the Commission on the completion of the internal market was the implantation of measures to create an area of free movement without border controls.

Frontiers in European law

Turning to the meaning of frontiers in European law – in its broadest sense as European Community law, European Union law and the European Conventions – one principle with special features can be distinguished.

The principle

As far as the principle is concerned, the classic concept of statehood and hence the concept of frontiers in international public law has not been changed by European Community law, nor by European Union law, nor by special European Conventions.

Modifications

However, when looking more closely at the role and meaning of frontiers in European law, there are two remarkable and perhaps modifying features in this law.

Community law contains the expression 'internal frontiers' in defining the internal market as 'an area without internal frontiers in which the free movement of goods, persons, services and capital is ensured in accordance with the provisions of this Treaty'.[2] European Union law develops this idea of classifying frontiers of Member States when stating that Member States shall regard as a matter of common interest 'rules governing the crossing by persons of the external borders of the Member States'.[3] This wording is used in the Schengen Agreements concluded between some of the Member States, and is used also in the Draft External Border Convention.

Consequences

The legal framework of the Community and the Union draws a distinction concerning the 'quality' of frontiers of the Member States by using the terms 'internal' frontiers and 'external' borders. It should be mentioned that the use of the two different terms 'frontier' and 'border' in the English version has no parallel in the German text, in which for both types of state boundary the term *'Grenze'* is applied.[4]

Different rules of European law are attached to the two frontiers: internal frontiers are not obstacles to the free movement of persons, goods, services and capital according to Community law, but the crossing by persons or goods of external borders is governed by different provisions.

European law plays a guiding role for both internal and external frontiers. Rules defining the meaning of internal frontiers are contained in EC law and in the Schengen Agreements signed by all but two Member States. Rules for external borders – the frontiers between Member States and non-Member States, and seaport and airport points of entry – can be developed in EC law and in the framework of the so-called Third Pillar of the Union (Co-operation in Justice and Home Affairs), and they are already contained in the Schengen Agreements.

Responsibility for frontiers?

Moving to the second question, 'Who is responsible for frontiers?', three connected subquestions must be distinguished: the legal foundations of responsibility; the concept of responsibility and the areas of responsibility; and responsibility in practice.

Legal foundations of responsibility

The idea of responsibility for frontiers is undefined. Legally, responsibility may be assigned by any legal order which defines duties and rights in relation to frontiers: in national law, in international law and in European law.

Concept and areas of responsibility

In general, the idea of responsibility concerns the attribution of consequences for an illegal act. The violation of rights and duties arising in the context of borders can be qualified as an illegal act. Responsibility constitutes a general principle of law within the meaning of Article 38 par. 1 lit. c of the Statute of the International Court of Justice. Thus the violation of a provision of public international law gives rise to an obligation on the part of the injuring state. However, both the conditions of responsibility and the consequences resulting therefrom remain vague and disputed. Attempts by the International Law Association to define more precise rules have failed so far.

Possible areas of responsibility in national law comprise the protection of frontiers against hostilities and criminal activities from outside, against the entry of people without permission or against the import of goods and capital. Responsibility for frontiers in international law may be conceived as the obligation to keep frontiers open and to ensure that people and territories outside the frontiers of a state are not harmed by any danger originating from within that state.

Responsibility in practice

Leaving aside responsibilities for the protection of frontiers in national law, the attribution of responsibilities appears both in a general international law context as well as in a special European law context.

Responsibility in general international law is particularly concerned with problems of outside effects: in particular for dangers emanating from within a frontier like radiating clouds from Chernobyl, invading troops from the former Soviet Union in Afghanistan, terrorists leaving countries of the Middle East or criminal groups coming, for example, from Romania.

In European law it is clear that special responsibilities are involved. Member States of the EC are responsible for state-caused impediments to the free movement of goods, persons, services and capital across internal borders.

The internal market aims at guaranteeing freedom of transfrontier private economic activities. This objective is frustrated by any openly discriminatory rules or practices, and by other unjustified impediments. Therefore Member States are obliged by rules of primary and secondary EC law to abstain from hindering free movement.

Only a few internal border controls on goods are allowed by the rules of Community law and decisions by the Court of Justice. Long before the Single European Act entered into force in 1987, the Court considered border controls to be an impediment to the free movement of goods and hence as a measure with equivalent effect to a quantitative restriction in the sense of the prohibition contained in Article 30 of the E(E)C Treaty. It allowed only a restricted number of justifications: namely controls necessary to meet the requirements of public interest as listed in Article 36 – controls necessary to levy internal taxes; controls necessary to check goods in transit; controls necessary to gain adequate data on intra-Community trade. The area without internal frontiers included in the E(E)C Treaty by the Single European Act tends to reduce even further cases of border controls which can be justified. As a consequence, controls on goods at internal frontiers have by now been widely abolished.

The internal market also included the free movement of persons and hence the abolition of respective border controls on those persons who enjoy the four freedoms. However, according to predominant opinion, Member States are not barred by Community law from controlling persons who cross an internal border as long as the Community has not ruled otherwise. The clear mutual obligation incurred by some Member States in the Schengen Agreement in 1985 to lift all controls of persons at common borders[5] goes beyond the present legal order of the EC.

If a Member State does not meet its obligations incurred under Community law, its responsibility to do so may be enforced in various ways. On the initiative of the Commission of the EC or of another Member State, the Court of Justice can rule that a Member State has failed to fulfil its obligations under the EC Treaty and shall be required to take the necessary measures to comply with the judgment of the Court.[6] On the initiative of an individual a Member State can also be held liable by a national court for damages caused by a qualified violation of Community law such as an unjustified impediment to the free movement of goods under certain limited circumstances.[7]

The second type of responsibility concerns the control of persons at external borders of the Member States. In this respect, Community law is not explicitly developed, except in relation to the powers of the Community to determine third countries whose nationals must be in possession of a visa when crossing the external borders of the Member States.[8] A regulation to this effect was issued by the Council in 1995. However, the rules governing the crossing by persons of the external borders of the Member States and the exercise of controls thereon are explicitly stated in the Treaty on European Union to be a matter of common interest.[9] An intergovernmental draft External Borders Convention was proposed in 1993, but is still not ratified. Outside both Community and Union law, the Schengen Agreements provide for controls of border crossing movement at the external borders by authorities of the contracting parties according to uniform principles (Article 6), including the principle of combating dangers to the national security and public order. Since these rules are outside the EC Treaty, no Community judicial procedures provided for by the EC Treaty are available nor is it possible to initiate a suit for recovery of damages if a Member State does not comply with its obligations. Nevertheless the contracting parties are bound by general international law.

Whose responsibility?

Turning to the third and last question, namely the question who is responsible, it is already evident that a distinction has to be made between the authorities responsible for the conduct at frontiers, and the fixing of criteria for the legally correct conduct.

Responsibility for conduct at frontiers

As far as the responsibility for conduct at frontiers is concerned, the basis of international public law has not been changed by European law. The Member State of the Union or of the Schengen area remains the competent authority, and therefore the State is responsible for the conduct of its agents at an internal frontier as well as at the external border. This is a valid principle as long as territorial authority and executive power remains with the Member States, and no executive authority is attributed to the Community.

Establishing criteria for proper conduct

Although criteria for the proper conduct of state authorities concerning frontiers rests in principle with the states, the criteria for both types of frontiers have been influenced by European law.

Policy for internal frontiers legally available to Member States of the European Union is restricted by rules of Community law. As seen above, Member States must not, in principle, systematically control goods at internal borders; they are also obliged, in principle, not to control persons entitled to free movement by Community law. The criteria of exemptions to that principle are governed by Community law, in particular by the *Cassis de Dijon* judgment and EC Treaty Articles 36 and 100a par. 4, for goods, and for persons by the interpretation of the provisions on the free movement of workers and of services, and on free establishment. Hence the sovereign exercise of internal border controls, although still available in principle, is barred by Community law. The result of this situation might be understood as a form of 'shared responsibility': while the Member States are responsible for guaranteeing the openness of internal frontiers (exemptions notwithstanding), Community institutions are responsible for interpreting the criteria of justified exemptions on the basis of EC primary law.

As far as the legally correct conduct of States at their external borders is concerned, this is influenced by several rules of European law related to the legal basis of responsibility.

European Community law

In the context of European Community law, national authorities have to comply with the binding effects of the Common Customs Tariff,[10] the Common Commercial Policy[11] and the Visa Policy[12] of the Community. In case of non-compliance, the usual legal remedies are available. In this respect, the sovereign exercise of external border controls is subject to rules of Community law. Again this situation might be seen as a kind of 'shared responsibility': while the Member States are responsible for policing borders with third countries, some criteria of the legally correct fulfilment of this task are not defined by States, but by a decision of the competent authorities of the Community.

European Union law

In the context of European Union law, Member States have to comply with those rules governing the crossing by persons of the external borders of the Member States and by the exercise of controls thereon contained in decisions taken or in conventions concluded in the framework of co-operation in justice and home affairs.[13] The Dublin Convention, for example, which was signed in 1990, may come into force; the crossing of the external border of a Member State by an asylum-seeker will then be a criteria for determining the State responsible for examining an asylum application.

European Conventions

In the Schengen Conventions, which have not been signed by all of the Member States of the Union, the conduct of the authorities of the contracting parties at their external borders conforms, according to the standards of international law, with the obligations incurred in those agreements. These obligations are specified in the Convention of Application of Schengen. The list of provisions includes that external borders may be crossed only at official points of passage and during the fixed opening hours; rules, including exemptions, and modalities of local border-crossing are promulgated by the Executive Committee (Article 3); the contracting parties guarantee that passengers arriving by plane from third States are controlled before boarding an internal flight as well as those travelling on an internal flight before boarding a plane bound to a third country (Article 4). Moreover, controls, though resting on national competence and responsibility, have to be performed according to certain uniform principles defined by the Convention (such as the extent of controls) and national law, and have to be performed taking into account the interests of all contracting parties (Article 6). This amounts to national responsibility for the common interest. In addition, the very detailed provisions on common visa policy towards third countries[14] contain elements of shared responsibility in so far as, for example, existing or developed common visa rules can be changed by unanimous agreement, and a uniform visa is valid for the territory of all contracting parties.

Perspectives

This overview shows that the traditional concept of national responsibility for frontiers has not disappeared, but has been partially modified by European law to include elements of national responsibility for the common interest or 'shared responsibility'. To meet certain defined needs, the addition of a Community responsibility, such as the establishment of a Community or Union Coast Guard to police the external borders, could be considered in the future.

Notes

1. Article 7a, EC Treaty.
2. Ibid.
3. Article K.1 (2), Treaty on European Union (TEU).
4. '*Binnengrenzen*' in Article 7a, EC Treaty; '*Außengrenzen*' in Article K.1 (2), TEU.
5. Article 2 par. 1.
6. Articles 169–171, EC Treaty.
7. See case *Brasserie du Pecheur*, reported in [1996] Vol. 1, *Common Market Law Reports*, 889 *et seq.* (Decision of 5 March 1996).

8. Article 100c par. 1, EC Treaty.
9. Article K.1 (2).
10. Article 18 *et seq.*, EC Treaty.
11. Article 113 *et seq.*, EC Treaty.
12. Article 100c, EC Treaty.
13. Article K.3, TEU.
14. Article 9 *et seq.*

Chapter 3

Boundaries and Boundary-consciousness: Politicizing Cultural Identity

ANTHONY P. COHEN

Boundary, culture, identity

My topic is concerned less with the political and jurisdictional aspects of the frontier than with people's consciousness of it.[1] Boundary is a term which is much more familiar to anthropologists than frontier. As I shall suggest later, it conveys a less-specific idea than 'frontier' – although anthropologists differ in this respect considerably from other social scientists. The proposition on which I wish to focus is that boundary is essentially a matter of consciousness and of experience, rather than of fact and law. As an item of consciousness, it is inherent in people's identity and is a predicate of their culture.

Apart from 'boundary' itself, the title of the present chapter includes within its brief span two frequently abused words, 'culture' and 'identity'. I shall attempt to be resolutely empirical. Without any semantic finesse, I shall treat identity as the way(s) in which a person is, or wishes to be known by certain others. 'Cultural identity' thus refers to the representation of the person or group in terms of a reified and/or emblematized culture. It is a political statement, present in those processes which we frequently describe as 'ethnic', and is composed of 'symbols'.

These are all words which have some currency in ordinary language, and whose academic and anthropological usage is thereby considerably complicated. In anthropology, 'culture' has gone through a succession of paradigm shifts. In the past it was used to suggest a determination of behaviour. There was also a major school of thought which treated culture as the means by which the supposedly discrete processes of social life, such as politics, economics, religion, kinship, were integrated in a manner which made them all logically consistent with each other. In this view, the individual became a mere replicate in miniature of the larger social and cultural entity. The tendency now is to treat culture much more loosely – as that which *aggregates* people and processes, rather than integrates them. It is an important distinction for it implies *difference* rather than similarity among people.

Thus, to talk about '*a* culture', is not to postulate a large number of people, all of whom are merely clones of each other and of some organizing principle. That is important for, in ordinary language, the word is still used all too frequently to imply this.

Moreover, if culture is not, *sui generis*, exercising a determining power over people, then it must be regarded as the product of something else: if not the logical replicate of other social processes – say, relations of production – then of social interaction itself. In this perspective, we have come to see culture as the outcome and product of interaction; or, to put it another way, to see people as active in the *creation* of culture, rather than passive in receiving it. If we are, in the contemporary jargon, the *agents* of culture's creation, then it follows that we can shape it to our will, depending on how ingenious and powerful we may be. And this, in the matter of the politicization of cultural identity, is another most significant characteristic to which we will return.

Culture, in this view, is the means by which we make meaning, and with which we make the world meaningful to ourselves. It is articulated by symbols. Symbols are quite simply carriers of meaning. To be effective, they should be imprecise, in order that the largest possible number of people can modulate a shared symbol to their own wills, to their own interpretive requirements: a tightly defined symbol is pretty useless as anything other than a purely formal sign.[2] Symbols are inherently meaningless, they are not lexical; they do not have a truth value. They are pragmatic devices which are invested with meaning through social process of one kind or another. They are potent resources in the arenas of politics and identity.

Culture, identity and symbolism all converge on the concept of ethnicity. In some respects, this is the most difficult word of the three, since it appears to mean something – indeed, has been imported into lay usage for this reason – but, in practice, means either everything or nothing at all. When a British politician or policeman says 'ethnic', they mean 'black' or, at least, 'different', 'other'. When the spokespersons of ethnic organizations say 'ethnic', they mean minority, usually disadvantaged or discriminated minority. When the racial theorist says 'ethnic', he refers to a relationship of blood and descent. If the word is to be anthropologically useful, it cannot refer exclusively to any of these. Ethnicity has become a mode of action and of representation: it refers to a decision people make to depict themselves or others symbolically as the bearers of a certain cultural identity. The symbols used for this purpose are almost invariably mundane items, drawn from everyday life, rather than from elaborate ceremonial or ritual occasions. Ethnicity has become the politicization of culture.[3] Thus it is, in part, a *claim* to a particular culture with all that that entails. But such claims are rarely neutral. The statement made in Ethiopia, 'I am Oromo', or in Northern Ireland, 'He's a Prod', are clearly not merely descriptive: they have an added value, either negative or positive, depending on who is speaking and to whom.

I have referred to the entailments of cultural claims. One aspect of the charged nature of cultural identity is that in claiming one, you do not merely associate yourself with a set of characteristics: you also dissociate yourself from others. This is not to say that contrast is the conscious motivation for such claims, as some writers have argued,[4] but it *is* implicit and is understood, the more so the more highly charged the situation may be. Cultural identity also entails a *patrimoine* and a history, or the acknowledged need to create these. It is in the expression of all of these entailments that the boundary, and especially the symbolic marking of the boundary, becomes crucial.

If the ethnic card is played in identity, it is not like announcing nationality. Ethnicity is not a juridical matter, carrying legal rights and obligations. It is a political claim, which entails political and moral rights and obligations. I use the word 'nationality' rather than 'nationhood' since, as we know, 'nationhood' may also be a statement of claim, and is one which is often made to emphasize the circumstances of its denial. The putative 'nationhood' of Scotland, or of 'the Jewish People' or of the Basques, are the axiomatic premises for claims: say *to* nationality, or to the legitimacy of Israel's occupation of so-called Judaea and Sumeria. But these are utterly different from the argument made by Hong Kong Chinese regarding their entitlement to a British passport; or from that of the British government concerning sovereignty over Gibraltar or the Malvinas. The one, nationality, is an argument about legal status. The other, nationhood, is a claim about the character and integrity of one's cultural identity. They may well coincide in a process which Løfgren describes as 'the nationalisation of culture'[5] in which attempts are made to forge a distinctive identity, for any of a variety of strategic reasons. His example is the creation of the national symbols and consciousness of 'Swedishness' in late nineteenth-century Sweden. But there is much other contemporary anthropological work on this issue, and the historian Peter Sahlins has ingeniously demonstrated the modulation over time of local and national identity in the transnational Pyrenees region of the Cerdagne (France)/Cerdanya (Spain).[6]

Now, the position has been taken in the past in anthropology that ethnicity – politicized cultural identity – was merely contrastive: that is, that it is invoked only to draw a real or conceptual boundary between one group and another. This suggests that the boundary is situational – invoked with respect to some groups and for some purposes, but not others. This does not sound much like 'frontier'. This position, associated originally with Fredrik Barth (1969), has dominated ethnicity studies for twenty-five years. It seems to me inherently unsatisfactory. Suffice it to say that in treating ethnicity just as a *tactical* identity, it ignored both *self*-consciousness and the *symbolic* expression of ethnic identity. The first suggests that an ethnic identity means different things to those who participate in it; the second directs us to the question of how ethnicity can have these infinitely variable meanings while still retaining its coherent expression.

Ethnicity – the political expression of cultural identity – has two distinctive registers to which we should attend. The first is used for the apparently dogmatic statement of more or less objective doctrine: 'I am a Palestinian' – and certain things will be understood as following from that. The second is for contentious statements which treat ethnicity as the context of, or as an aspect of identity with very uncertain implications: 'I am a *particular* Palestinian'. The apparently monolithic character of ethnic identity at the collective level thus does not pre-empt the continual reconstruction of ethnicity at the personal level. Ethnicity has a definite appearance, but rather indefinite substance.

This same discrimination of appearance from reality, of substance from insubstantiality, is pertinent to the related idea of 'boundary'. This most topical of terms, or the entity which it expresses, seems to have preoccupied the social sciences since the late 1980s and the collapse of the central European state socialist empires. In the attempt to shed some conceptual light on a categorical morass, Malcolm Anderson distinguishes among 'frontier', 'boundary' and 'border':[7]

> 'Frontier' is the word with the widest meaning. . . . In contemporary usage, it can mean the precise line at which jurisdictions meet, usually demarcated and controlled by customs, police and military personnel. 'Frontier' can also refer to a region. . . . Even more broadly, 'frontier' is used in specific cases to refer to the moving zone of settlement in the interior of a continent. . . . The term 'border' can be applied to a zone, usually a narrow one, or it can be the line of demarcation. . . . The word 'boundary' is always used to refer to the line of delimitation or demarcation and is thus the narrowest of the three terms.[8]

His usage is similar to that proposed by Coakley:

> Political geographers conventionally distinguish between *boundaries*, which have a precise, linear quality, and *frontiers*, which have more diffuse, zonal connotations. The concept of frontier has a broader social significance than the more restrictive legal concept of boundary.[9]

The confusions among these words, all of which express the condition of contiguity, are those of ordinary usage rather than of science. It might be helpful to think less in terms of discriminating among them on the grounds of their putative referents – since ordinary language will not honour such precision – than in terms of how they are used and what they are used for. In the discourse of anthropology, such a taxonomy of concepts and attitudes (rather than of concrete empirical referents) would suggest almost the opposite of Anderson's (1996) and Coakley's (1982) surveys: that 'boundary' is the word with the most general application (since, in anthropology, it has been used to signify such diverse things); whereas 'border' seems situationally specific, and 'frontier' has come to be reserved to fairly strictly limited geopolitical and legal applications. Thus, Piero Vereni makes the frontier the line which asserts the difference between neighbouring national groups, and which is then affirmed (or otherwise) by the achieved 'fact' of a border.[10] Generally in anthropology the distinction can be accomplished simply by regarding frontiers and borders as matters of fact; whereas boundaries are the subjects of claim based on a perception by at least one of the parties of certain features which distinguish it from others. Whether it refers to a collective condition, such as ethnic group identity, or to something as ephemeral as 'personal space', boundary suggests contestability, and is predicated on consciousness of a diacritical property.[11]

There is a tendency among anthropologists (and, indeed, among other social scientists who write about ethnicity) to credit the concept of boundary to Barth's (1969) seminal symposium, *Ethnic Groups and Boundaries*; and, by implication, to associate it with ethnicity (or, as the subtitle of Barth's book put it, with the social organization of cultural difference). But the concept is really much more fundamental to the discipline and to the nature of our enquiry. When anthropologists defined the subject as the study of other cultures, they necessarily (if unwittingly) placed 'boundary' at the very centre of their concerns. The relativism of anthropologist/anthropologized, us/them, self/other, clearly implies boundary.

The problem became fixed as one inhering in the distance between *cultures* rather than between *minds*. Anthropology has been preoccupied with the boundaries between cultures. It has preferred to avoid the boundaries between minds, between consciousnesses, either because these have been regarded as too difficult to cross,[12] or because such a refocusing of enquiry would have subverted the disciplinary practice of generalization and its conceptual bases. This more fundamental problem has been shoved aside simply by predicating consciousness on culture, which is itself anthropologically constructed as being different from, and therefore 'relative to', *other* cultures.

One consequence of this has been that anthropologists have been largely content to assume the existence and integrity of collective boundaries. Rather than questioning their existence, or questioning the extent to which they might reasonably be generalized (*whose* boundaries are they?) they have been concerned almost exclusively with the ways in which boundaries are marked. There have been significant theoretical debates concerning the differences among the ways in which they have done this, and concerning the nature of the boundary-marking devices and processes which they have attributed to people. But there is little room for doubt that their concern has not extended to the more fundamental question. It has been so central an ethnographic preoccupation that examples would be somewhat gratuitous, but to give just an idea of their range: it could be found among Leach's 'aesthetic frills', those non-technical aspects of ritual which express collective identity by emphasizing cultural possession (see above). It was explicitly at the heart of Schwartz's depiction of the 'ethnognomomic' activities of Admiralty Islanders.[13] It was strikingly and movingly present in Eidheim's famous account of the reaction of Norwegian Saami to the stigma they supposedly perceived as attaching to them;[14] and provided the material for the reformulation of migrant West Indian identities among the Notting Hill carnival participants described by Abner Cohen.[15]

So ubiquitous has this kind of work been, especially in studies of ethnicity and social identity, that we have taken for granted the integrity of its central concerns: to show how individuals are constructed in the images of their collective representations. It has imputed boundary-consciousness to people without pausing to enquire quite what it is that they are supposed to be conscious of. What is the individual conscious of when she or he invokes a boundary as a means or source of social identity? Culture and consciousness are not alternative modalities: culture only exists *as* consciousness. In the ethnographic literature, people have been constructed in terms of putative boundaries (localities), and in terms of anthropologists' consciousness of boundaries, without adequately interrogating these notions.

The terms 'frontier' and 'border' (and boundary, if it is not distinguished from them) alerts us to lines which mark the extent of contiguous societies, or to meeting points between supposedly discrete social groups. We have barely glanced at those more amorphous divisions which appear routinely, not just between cultures nor even within them, but between intimates who share culture. As I have suggested, we have shied away from, have even denied any interest in, the boundedness of the mind, the limits of consciousness which separate one self from another. We have excused ourselves from such an enquiry on the grounds that it would be too difficult, and that our concept of culture enables us to invent people who are similar to each other. Instead of dealing with the individual, we have restrained our ambition and addressed ourselves instead to whole societies or to substantial parts of them. Yet, looking at individuals' boundary transformations should alert us to the qualitative nature of *collective* boundaries.

I will try to illustrate this briefly with reference to rituals of initiation. In dealing with ritualized status passage, we do not seem commonly to have explicitly applied the concept of boundary to divisions between statuses, but there is no reason why we should not do so. We have the evocative notion of liminality to describe the blurriness of transformation, and the acute consciousness of status on either side of it. This is not unlike the exaggerated concern with social identity to be found commonly in geopolitical borderlands,[16] and to which I shall return. But the difficulties of passing from status to status seem curiously understated in ethnographic accounts – as if such adjustments were as unambiguous as (albeit more troublesome than) crossing a national border: one moment you are in Italy, the next in France. The worst you are likely to suffer is a brief spell in no-man's land. So it is with accounts of initiation. One day the initiate is a child; the next he or she is initiated and, after due process of seclusion, re-emerges into society bearing the new status of adult, or initiated youth, or marriageable girl. The confusion of liminality, the blurriness of being 'betwixt and between', or being in the social equivalent of no-man's land, is somehow confined temporally by the ritual process and spatially to the initiates' lodge. It is ended by the next ritual phase of reaggregation. This seems hardly plausible to me. Transformations of status, like crossing geopolitical borders, require a process of adjustment, of rethinking, which goes beyond the didactic procedures about which we have been told so much. They require a reformulation of self which is more fundamental than admission to items of lore, or being loaded with new rights and obligations. The difficulties inherent in such self-adjustment may vary according to the nature of the frontiers which are crossed; but our experience of politics and travel should also alert us to the deceptively innocuous character of crossing between supposedly proximate statuses or cultures. The first intimation to us that we are *really* in a different place may be the look of incomprehension on the faces of our interlocutors, or the pained censure by others of our newly inappropriate behaviour. Having crossed a boundary, we have to think ourselves into our transformed identity which is far more subtle, far more individualized than its predication on status.

A boundary-crossing stimulates the awareness of a person *as an individual*, as someone who can step back and reflect on his or her position with respect to society. If we recognize boundaries as matters of consciousness rather than of institutional dictation, we see them as much more amorphous, much more ambiguous than we otherwise have done. It may be this very ambiguity which inclines societies to invest their various boundaries so heavily with symbolism. The contributors to my symposium, *Symbolising Boundaries*, all describe such processes of marking in the undramatic circumstances of the British Isles, whether dealing with the imagery of suburban Manchester or with adolescence in rural Northumberland.[17] As a matter of ideology, the boundary may be given dogmatic form. But its internalization in the consciousness of individuals renders it much less definitely. I think perhaps this offers us a clue to the discrimination of boundary, border and frontier. Border and frontier have the quality of finity, definity, about them. When they are crossed, one has definitely moved from the Cerdagne to Cerdanya. That is undeniable, for my passport stamp tells me so. What is much less certain is what this crossing-point means to those who live on either side of it. The uncertainty may be glossed by language, currency, by law, lore and by all the iconography of custom and tradition. But, when all this is said and done, it remains a gloss on the much more ambiguous boundaries of consciousness. Borders and frontiers seem to me to have something in common with the taxonomic absoluteness of anthropological categories; boundaries, with the blurriness and elusiveness of symbols.

Of course it follows that if one does not know quite what it is that has been crossed, then one may also be unaware *that* a boundary has been crossed at all. As an English person resident in Scotland, I feel an intolerant dismay at the insensitivity of many incoming English people to the notion that in Scotland they are actually in a different place. No doubt we all have examples from our field experience of people who fall over their idiomatic feet because of their cultural boundary errors, and this kind of insensitivity or clumsiness is also readily observable among those crossing unfamiliar status boundaries. Again, the examples are legion and perhaps do not need to be cited to make the elementary proposition that, as objective referents of meaning (as opposed to political legitimacy) boundaries are essentially contestable, while borders are not.

In an intriguing examination of Canadian fiction writing, Russell Brown has showed how central the border is to Canadian identity. Actually, his claim is more ambitious: 'the border is central to Canada's *self-awareness*'.[18] Without wishing to be pedantic, I find difficulties in this claim: countries are not self-aware, people are. If he is saying that the border is significant in individuals' awareness of themselves as being Canadian, that is fine. But if he is saying that in so far as individuals are aware of themselves it is *as* Canadians because the border looms so large, I would have to regard this with some scepticism; and as a failure to appreciate the complexity of self-identity. He points to the ubiquity of oppositions as a theme in Canadian literature, but it does not need a structuralist to point out that there is nothing peculiarly Canadian about this. Any anthropologist with experience of peripheral societies, or of societies in which boundaries are heavily invested symbolically, would have made similar observations – but not because of the *border*: the border is a social fact. Whether or not it signifies difference is a matter of social construction, and is more properly thought of as one of *boundary*. If border is fact, boundary is consciousness, and the difference between them is crucial. I suggest a distinction, the significance of which I can only assert but not demonstrate. There is a difference between being conscious of what is on either side of a border, and being preoccupied

with the boundary as such. The first, again, implies definity: if I go this way,
I will be X; if I go that way, Y. The latter seems to me more authentically
boundary-conscious: liminal, uncertain, unpredictable.

It is this kind of uncertainty which drives people to grasp for certainty, and
which in turn motivates identity. This may be formulated around a collective
stereotype or dogma, such as 'Canada's self-awareness'. Or, it may proceed
the other way around, by assimilating such cultural products to self-
experience. Writing with respect to the Cerdagne/Cerdanya border, and
following Benedict Anderson, Sahlins says that national identity,

> appeared less as a result of state intentions than from the local process of
> adopting and appropriating the nation without abandoning local interests, a
> local sense of place, or a local identity.[19]

Historically, anthropology has privileged the collective and dogmatic, and
neglected the individual and experiential, as a consequence of its general
neglect of selfhood and self-consciousness. It is a neglect which requires
repair if we are really to get to grips with the meanings of boundaries.

Ethnicity, nationhood, boundary-consciousness

If we return to ethnicity, we will find that a focus on boundary-consciousness
will sensitize us to the kinds of circumstance in which ethnic identity
becomes salient, in which people's consciousness of themselves *as* ethnic
becomes prominent. The minimal conditions are that people recognize that
ignorance of their culture among others acts to their detriment; that they
experience the marginalization of their culture, and their relative power-
lessness with respect to the marginalizers.[20]

With ignorance of a culture goes the denial of its integrity. Because culture
is expressed symbolically, and thus has no fixed meanings, it is often invisible
to others, especially to powerful others. This denial of, or threat to, cultural
integrity is experienced by people in all manner of ways: through the
subordination of indigenous languages – say, Tamil to Sinhala; Breton to
French; Catalan and Gallego to Spanish; Welsh and Gaelic (like French,
among Quebecois) to English; through the denigration of their tradition (the
examples are almost limitless – Australian Aborigines; Mongolian Buryats;
Basques); and from the outright denial of their distinctiveness – say, Armen-
ians and most other nationalities in the former Soviet Union, sectarian
groups in South Asia, and so on.

It does happen, *has* happened historically on a massive scale, that such
continuous denigration seems to drive people into cultural retreat where
they either make their tradition a *covert* matter, or appear to desert it in large
measure. But the historical era in which this retreatist stance prevailed came
to an end emphatically during the later 1960s, and was replaced by an

assertive stance in which the putative stigma of cultural inferiority was transformed into an emblem of its superiority.

So, from the experiential point of view, the politicization of cultural identity requires people to react against their own felt disadvantage and denigration, as well as occurring in characteristic economic and political circumstances.

Ethnicity is also a matter of historical genre, and we seem frequently now to find ourselves in the very middle of its often grim apotheosis. The process of decolonization appears to have no objective end to it: its logic is to continue the process to a kind of infinite federalism. Pierre van den Berghe describes ethnicity in the industrialized world as 'the last phase of imperial disintegration'. He asks, 'if the Fiji Islands can be independent, why not Scotland?'[21] If Sri Lanka, why not a Tamil state within the island; if Ethiopia, why not Eritrea (again)? And so on. Almost everywhere one turns, there is being played out an epic struggle for recognition, for the acknowledgement of rights, above all, for the acknowledgement of cultural integrity, expressed now in terms of nationhood, even of statehood. Nationalism now seems more a matter of iconicity than of political economy; and, as such, is being pressed by nationalist activists onto the consciousnesses of ethnic group members. In this militant mode, there is the attempt to bring people's boundary-consciousness into alignment with putative political frontiers.

The imperative need to posit culture as identity and, increasingly, as ethnicized national identity, can arise from many different circumstances. I have mentioned earlier those of a perception of imminent and possibly cataclysmic crisis; and of the attempt to reverse extreme disadvantage. It appears also when there is a perceived threat to the distinctiveness of a group through its assimilation or the blurring of its boundaries, or as the consequence of internal differentiation or disagreement. One finds then a politicization of culture or tradition or whatever putative dogma provides the *raison d'être* of the group. The call for a *jihad* to unite nations against a common enemy, in order to mask the internecine nature of dispute among them; the spurious elevation of Zionism to the status of religious obligation; and the 'metaphorization' of culture as a response to such historical circumstances as demographic and economic change, secularization, integration and vulnerability to new kinds of information.

Ethnicity, nationhood, 'peoplehood', came to be contrived using symbols which can be made by individuals to mean anything, to encompass widely varying kinds of personal experience and material conditions. It would be an unjustifiable generalization to construe as cynical these representations of identity in somewhat contrived cultural terms: their expression and use may well speak rather of a commitment to the integrity of culture and group. In these cases, it is only by making the culture visible, so to speak, that its bearers can gain some awareness of what they have to defend, and those to whom it is vulnerable can be made aware of what they might otherwise damage, unwittingly or deliberately. So far as indigenes are concerned, the iconization of culture may be no more than a means of agreeing on a very limited number and range of symbols as a kind of lowest common denominator, which can be interpreted and rendered privately by each of them in ways to suit themselves. Apparent uniformity in the terms of public discourse glosses over an uncountable multitude of divergences of meaning. But for outsiders, it is a caution against their cultural blindness.

For example, the tradition of the Fourth World peoples frequently refers to their incomparably expert use of the land, as either hunters and gatherers, or as pastoralists, which was based on their expert knowledge, and attitudes to the land, both of which are peculiar to the people themselves and are not accessible to outsiders. The reason is quite simple: outsiders cannot see the land in the same way; and, therefore, they cannot know what they are looking at. The cultural boundedness of Australian Aboriginal perception has been amply documented both in ethnography and in fiction. The stereotypical insensitivity of the white to Aboriginal sacred places may be a consequence of contempt in some cases; but is certainly the result of cultural blindness in most. To us a stone is a stone; to the Aboriginal, it conceals an ancestor. Another celebrated example is the blindness of Canadian government scientists to the superior indigenous expertise of Mistassini-Cree Indians in Quebec in monitoring the environment, an expertise underpinned by their cultural rights and obligations with respect to the use and stewardship of land.

Wherever one sees this kind of struggle – whether in Norwegian Saamiland; in the Torres Straits Islands; among Kayapó Indians; in South Asian 'communalism' or in Southern Africa – there seems to be an almost irresistible inclination to explain behaviour by treating it as the product of culture: the Zulus or Yanomamo are said to be warlike or aggressive; some other society might be spoken of as constrained in its thought and action by its cosmology or its kinship system or whatever. There is a fundamental confusion here between culture as a body of substantive fact (which it is not) and as a body of symbolic form which provides means of expression but does not dictate what is expressed or the meaning of what is expressed. In this respect, culture is insubstantial: searching for it is like chasing shadows. It is not so much that it does not exist, as that it has no ontology: it does not exist apart from what people are conscious of, apart from what they *do*, and therefore what people do cannot be explained as its product. Culture can be invoked as a means of representing them – as, for example, when it is deployed as identity. But in those circumstances it must be regarded in the same way as any other symbolic expression: as being inherently meaningless, but capable of substantiation at the discretion of those who use it – multireferential, multivocal, an infinitely variable tool. It is the consequent diversity of meaning which requires us to make a clear distinction between boundaries in people's consciousness, and the legal representation of their distinction from others through borders or frontiers.

It does not matter whether we use the anthropologists', or political scientists' and geographers' taxonomies of border concepts, so long as we do make the distinction between barriers in jurisdictional fact and in the mind. We must do this in order not to fall prey to the comfortable assumption that the nationalities on either side of an international dividing line are co-extensive with discrete cultures which themselves dominate and are replicated through the behaviour of individuals. The meaning of the division is to be sought in the consciousness of those who are oriented to it, not in some abstracted collectivity. Unless we recognize the power and persuasiveness of such boundary consciousness, we cannot begin to understand the attraction to people of ethnicity, nationhood or any other collectivity which claims distinctiveness for itself.

National identity and Europeanness

With regard to the specific focus of this book, it seems to me that the pertinence of this kind of argument for the study of frontiers in Europe is, first, to call into question the generalizability of what the frontier may signify on either of its sides; and, by the same token – and, perhaps, even more obviously – the meaning of Europeanness. The issue has come to be politically pressing for at least two reasons: globalization and migration. Globalization and, what Hannerz has identified as its correlate, creolization,[22] has changed, but has not diminished – indeed, may have emphasized – the significance of cultural or jurisdictional barriers. Nationalist sentiment and its political expression remains rampant. Moreover, as Malcolm Anderson has shown in his new book,[23] the permeability of jurisdictional frontiers by the internet or by old-fashioned radio has called forth new jurisdictional devices. Migration, temporary or permanent, suggests that cultural boundaries and jurisdictional frontiers are not coincident, their disjunction producing racist effusions throughout Western Europe.

The mutual implication of jurisdictional frontier and interpretable, meaningful boundary is a matter which has bypassed the so-called Eurosceptics in the British Conservative Party. There may be plenty to object to in the Common Fisheries Policy or, for that matter, in the fabled single currency. But I very much doubt whether either of them necessarily entails the loss of national identity. To the contrary, if I am correct, they will intensify boundary-consciousness.

Notes

1. This paper incorporates an earlier essay published in *Revista de Antropología Social*, 3, 1994, pp. 49–61. I am most grateful to the editor, Professor Ricardo Sanmartín Arce, for his permission to use it here. It also draws heavily on another paper, 'Boundaries of Consciousness, Consciousness of Boundaries', in H. Vermeulen and C. Govers (eds) *The Anthropology of Ethnicity: Beyond 'Ethnic Groups and Boundaries'*, Amsterdam: Het Spinhuis, 1994.
2. Cohen (1985), p. 18.
3. Paine (1984), p. 212.
4. Barth (1969), Boon (1982).
5. Løfgren (1989).
6. Sahlins (1989).
7. His dismissal of the term 'march' as 'archaic' (referring to the outer limits of a given territory) will not satisfy anthropologists for whom it has been modernized into the troublesome 'margin'.
8. Anderson (1996), p. 9.
9. Coakley (1982), p. 36.
10. Vereni (1996), p. 82.
11. In her study of the annual festival in a Scottish Borders community, the geographer Susan Smith (1993) conflates all three terms by making them expressive of 'space' (a word to which human geographers seem to resort much as anthropologists do to 'culture'). While there may be no intrinsic value in discriminating the three words, there surely is something to be gained from distinguishing the material from the ideal.
12. e.g. Needham (1981).
13. Schwartz (1975).
14. Eidheim (1969).
15. Cohen, Abner (1980, 1992).
16. e.g. Sahlins (1989), Brown (1990).
17. Cohen (ed.) (1986).
18. Brown, R. (1990), p. 32.
19. Sahlins (1989), p. 9; see also ibid., pp. 269–70.
20. Cohen (1975).
21. van den Berghe (1976).
22. Hannerz (1987).
23. Anderson (1996).

Bibliography

Anderson, M. (1996) *Frontiers: Territory and State Formation in the Modern World*, Oxford: Polity.

Barth, F. (1969) 'Introduction', in F. Barth (ed.) *Ethnic Groups and Boundaries: the Social Organization of Culture Difference*, London: Allen & Unwin, pp. 9–38.

Berghe, P. van den (1976) 'Ethnic Pluralism in Industrial Societies: a Special Case?', *Ethnicity*, **3**, pp. 242–55.

Boon, J. A. (1982) *Other Tribes, Other Scribes: Symbolic Anthropology in the Comparative*

Study of Cultures, Histories, Religions and Texts, Cambridge: Cambridge University Press.

Brown, R. (1990) *Borderlines and Borderlands in English Canada: the Written Line*, Borderlands Monograph Series 4, University of Maine.

Coakley, J. (1982) 'Political Territories and Cultural Frontiers: Conflicts of Principle in the Formation of States in Europe', *West European Politics*, **5**(4), pp. 34–49.

Cohen, Abner (1980), 'Drama and Politics in the Development of a London Carnival', *MAN* (n.s.), **15**, pp. 65–87.

Cohen, Abner (1992) *Masquerade Politics: Explorations in the Structure of Urban Cultural Movements*, Oxford: Berg.

Cohen, A. P. (1975) *The Management of Myths: the Legitimation of Political Change in a Newfoundland Community*, St John's (Newfoundland): ISER.

Cohen, A. P. (1985) *The Symbolic Construction of Community*, London: Routledge.

Cohen, A. P. (ed.) (1986) *Symbolising Boundaries: Identity and Diversity in British Cultures*, Manchester: Manchester University Press.

Cohen, A. P. (1994) *Self Consciousness: an Alternative Anthropology of Identity*, London: Routledge.

Eidheim, H. (1969) 'When Ethnic Identity is a Social Stigma', in F. Barth (ed.) *Ethnic Groups and Boundaries: the Social Organization of Culture Difference*, London: Allen & Unwin, pp. 39–57.

Hannerz, U. (1987) 'The World in Creolisation', *Africa*, **57**, pp. 546–59.

Løfgren, O. (1989) 'The Nationalization of Culture', *Ethnologia Europaea*, **19**, pp. 5–23.

Needham, R. (1981) 'Inner States as Universals', in *Circumstantial Deliveries*, Berkeley: University of California Press, pp. 171–88.

Paine, R. P. B. (1984) 'Norwegians and Saami: Nation-State and Fourth World', in G. L. Gold (ed.) *Minorities and Mother-Country Imagery*, St John's: ISER.

Sahlins, P. (1989) *Boundaries: the Making of France and Spain in the Pyrenees*, Berkeley: University of California Press.

Schwartz, T. (1975) 'Cultural totemism: ethnic identity, primitive and modern', in G. de Vos and L. Romanucci-Ross (eds) *Ethnic Identity. Cultural Continuities and Change*, Palo Alto: Mayfield, pp. 106–31.

Smith, S. J. (1993) 'Bounding the Borders: Claiming Space and Making Place in Rural Scotland', *Transactions of the Institute of British Geographers* (n.s.), **18**, pp. 291–308.

Vereni, Piero (1996) 'Boundaries, Frontiers, Persons, Individuals: Questioning "Identity" at National Borders', *Europaea*, **II–1**, pp. 77–89.

Chapter 4

Sovereignty, Citizenship and Nationality: Reflections on the Case of Germany

JOHN BREUILLY

The modernity of the national frontier

The process of nation-state formation in modern Europe is closely associated with the increasing importance of national frontiers as sharp lines of division between various states. There are two aspects to this.

First, various functions requiring the surveillance and control of frontiers had tended, before the modern period, to be linked to different boundaries. Poor relief operated within local boundaries; the levying of tariffs within regional boundaries; military security within state frontiers. Furthermore, the implementation of many of these functions did not require that a sharp boundary be defined but operated instead within a zone. In the modern nation-state all these various functions have come to be linked to the surveillance and control of one frontier – that of the national territory – and this as a clear demarcation line, not a zone.

Second, most of these boundaries or zones were regarded as little more than functional. Military border areas, municipal or parish boundaries, the tariff-free areas of particular provinces or even several states grouped into a customs union (such as the German *Zollverein*) – none of these frontiers were vested with a symbolic meaning over and above the mundane functions with which they were associated. 'Sacred' frontiers, where claims to such were made, were often quite different from mundane ones (for example, the Rhine or the Alps or the Pyrenees). The formation of nation-states brought with it the idea that the national territory and, by implication, the frontiers demarcating that territory from other states, was more than a geographical area in which certain institutions functioned but also had a sacred meaning. This kind of meaning is often neglected in social scientific treatments of frontiers as well as making life difficult for pragmatic diplomats.[1] How and why did so many different functions and meanings come to concentrate themselves upon the linear frontier of the national state? What is the relationship between the convergence of different functional meanings and the acquisition of a sacred significance?

This impressionistic essay will address such questions to Germany. In this case the national territory, however variously defined, was far larger than any one state before unification in 1871. Whereas in Spain, France and Britain nation-state formation only involved 'nationalizing' the frontiers of the existing state, in cases such as Italy and Germany it involved the destruction of existing state frontiers and the construction of new, larger nation-state frontiers. A third type of case involved the destruction of existing state frontiers and the construction of smaller nation-state frontiers in their place.[2]

This essay will begin by considering the meaning state and national frontiers had within nineteenth-century German states before the formation of a German nation-state in 1871 and how far a 'nationalizing' process can be observed at work in that earlier period. That in turn will set the scene for considering the changing meanings the national frontier acquired after the formation of a German nation-state. The essay will end with some comparisons between the German frontiers and other national frontiers and between the role of frontiers in the nineteenth-century 'German political system' and the present 'European political system'.

Frontiers and citizenship

A great deal has been written about the ideological meanings attached to German frontiers; it is part of the story of German nationalism.[3] However, apart from being a well-known story, this only tells us about the preoccupations of nationalist minorities. The concern here is rather with the significance state and national frontiers had for the majority of Germans. My assumption is that for this majority nationalist ideology meant little until well after the formation of the nation-state.[4] However, that does not mean that frontiers of various kinds, including national ones, did not matter a good deal to them for other reasons. I will explore these by relating frontiers to citizenship. First it is necessary to sketch in some points about the development of modern citizenship and how this is connected to frontiers.

T. H. Marshall distinguished three kinds of citizenship: legal, political and social.[5] Legal citizenship means equality before the law. Political citizenship means rights of participation in affairs of government, above all electing representatives and being able to stand for election. Social citizenship means right of access to social goods, such as food, clothes and housing, health-care and education, when unable to make such provision for oneself. Marshall used these three elements of citizenship to construct a view of modern British history, suggesting that the achievement of citizenship proceeded sequentially. The conflicts over civil and religious liberty between the mid-seventeenth and the early nineteenth century achieved legal citizenship;

those over the extension of the parliamentary franchise and the supremacy of the House of Commons secured political citizenship during the nineteenth and early twentieth centuries; those over the construction of a welfare state in the twentieth century produced social citizenship.

I want to use Marshall's broad scheme for a rather different purpose. I suggest that some functions associated with frontiers are related to the evolution of different citizenship rights. These I slightly rephrase from the terms employed by Marshall as: rights to legal equality in matters concerning property and person; rights to political participation; welfare rights, above all poor relief. There are also modifications that need to be made to Marshall's scheme. First, even if one accepts the sequence he posits for the British case, it does not work for other cases. Indeed, it could be argued in the German case that the whole process was telescoped into a much briefer period, with forms of legal, political and social citizenship all being achieved over the same period during which a German nation-state was created and consolidated, creating tensions and 'deficiencies' which were absent from the more gradualist British model.[6] Second, citizenship involves obligations as well as rights and I will consider in particular those associated with taxation and military service. Third, Marshall, writing about Britain, could take as fixed the territorial framework within which citizenship rights were achieved. This was not the case in much of Europe. Here disputes over citizenship were bound up with conflicts over frontiers. These conflicts took the form not just of internal political struggles, including civil war, but also of war between states. In such intense conflicts frontiers did not simply demarcate the area within which certain functions were performed but also became symbolic matters which were bound up with the very power and identity of the state and its citizens. This in turn stresses the close links between nationality and citizenship, between state frontiers as functional and as sacred. With these points in mind, let us turn to the case of Germany.

Germany before 1871 as a multi-state system

Germany was divided into many different states before 1871: hundreds before Napoleonic conquest and reorganization; some thirty-nine between 1814 and 1866; six between 1867 and 1871; and two after 1871.[7] This means that the first phase in the development of modern citizenship took place within individual German states. In this process the frontiers dividing those states from each other mattered a good deal. Yet the German states also formed a loose system which distinguished them from non-German states.

Frontiers and citizenship rights

Legal citizenship

Ancien régime European society was organized on the basis of privilege. In such a society boundaries meant different things to members of the different estates. An individual might owe different allegiances to a lord, a church, a prince and a town. No single boundary enclosed these different authorities. Absolutism to some extent simplified matters by overriding many of the powers attached to privilege, but it tended to do so in relation to rights of participation at the centre rather than local social and economic distinctions.

The French Revolution brought about radical change. The attack upon privilege led to sweeping assertions of legal equality, expressed in the codification of civil law. Fear induced through war with much of Europe led to the drawing of sharp distinctions between French citizens and aliens. Suddenly the French nation was defined as a body of citizens, legally distinguished from other nations, with national frontiers clearly demarcated.

French military success in the German lands meant that radical changes were also made there. Many smaller states were merged to make bigger ones and privileged orders were swept away. This altered the forms and functions of many boundaries. The abolition of guilds meant the abolition of the monopoly power such guilds had exercised within a defined locale. The abolition of urban/rural distinctions was linked to the imposition of a common administrative and fiscal system. The French often highlighted the radical nature of these changes by allocating completely new boundaries to local and regional government, on the same lines as it had with its own departments and cantons. The result was a radical redrawing of the political map of Germany, both in terms of states and sub-state frontiers, associated in turn with the construction of a direct state–subject relationship within these frontiers replacing a series of regime–privileged estate relationships operated within different and varying frontiers.[8] Some geographers, faced both with constant changes to political frontiers and an emphasis on boundaries as rational, man-made features began to develop ideas of natural or rational frontiers to replace those based on tradition and sentiments.[9]

Most of these innovations were not reversed after 1815 and, even where they were, the reversals were often little more than cosmetic. The post-1815 German states, often a continuation of Napoleonic creations, wished to assert their sovereignty against church, nobility and town. State officials energetically imposed legal equality upon an often reluctant population, especially those for whom it meant the removal of privilege.

Nevertheless, these were policies pursued by individual states. Over and above these states the peace settlement concluded at the Congress of Vienna in 1814–15 had created a 'national' political system in the form of the German Confederation (*Deutscher Bund*). Article 18 of the Federal Act of 1815, the 'constitution' of the *Deutsche Bund* had sketched out some common legal rights to do with property and freedom of movement. However, little practical progress was achieved along these lines.[10] Even within individual states there were variations in legal rights between different regions – for example, there was much more legal equality in the Rhineland province of Prussia which inherited many French legal reforms than in the eastern provinces. Indeed, the most effective action taken by the *Bund* involved repressing civil rights by applying restrictions on freedom of speech, publication, assembly and organization to all the member states.

The first clear link between German citizenship and legal equality came with the revolution of 1848 and the convening of a German National Assembly (henceforth, the GNA). The GNA in the constitution it drew up in 1848–9 decreed legal equality. It proposed German citizenship rights which would enable people to move freely across state frontiers, along with rights of residence, pursuit of an occupation and acquisition of property.[11]

The failure of the revolution meant that this constitution was never implemented. Advance towards equal legal rights across state frontiers came about in three other ways up to 1866. First, there was the *Zollverein*. This was only a customs union but it did enable citizens of the Member States to trade freely across state frontiers. Second, there were various agreements on population movement. In the 1820s these took the form of bilateral agreements concerned with deportation. They were followed in the 1850s and early 1860s by multilateral agreements (achieved by states affiliating to a treaty drawn up by Prussia) on the rights of subjects to move across state frontiers and to acquire citizenship in the state to which they had emigrated.[12] Third, there were some efforts to create common rules in such areas as commercial law by the *Bund*.[13] Each of these arrangements reflected some functional concern about the movement of goods or people across state frontiers. Boundaries were set aside for particular pragmatic reasons. However, 'national' legal rights of this kind were only a small part of the general achievement of legal equality. For most people this largely took place within individual German states.

Political citizenship

Legal equality was imposed from above by the state. It was not necessarily accompanied by the creation or extension of political citizenship. Indeed some people, usually enlightened state officials, argued that the establishment of powerful parliaments with large electorates would lead away from legal and economic liberalism as different groups exploited such political institutions to buttress and protect their particular interests. Some German states issued constitutions soon after 1815 but this was more a matter of asserting a common authority over their various, often newly acquired territories than a move towards power-sharing. For most people the most important participatory institutions, where these did exist, were at the local level, for example municipal councils. Nevertheless, in some south German states the creation of statewide parliaments with consultative rights helped in the formation of a statewide public opinion and political opposition.

There was a temporary breakthrough to political citizenship in 1848, most dramatically symbolized when millions of Germans (adult males) took part in the election of a German parliament. But the moment was short-lived and did not lead on to the development of a national parliament, regular elections and the formation of political parties at a national level.

However, the counter-revolution did not simply restore the *status quo ante*. Prussia introduced a constitution in December 1848 which included a parliament with a lower house elected on the basis of universal and equal manhood suffrage. The equality was soon removed in 1850 with the introduction of the three-class franchise but now, at an all-state level, there was a parliament to set alongside the dynasty, and its instruments, the civil service and army. The law establishing the constitution was the first official document to refer to the Prussian state in the singular instead of the plural. In 1848 Prussia officially ceased to be a plurality of provinces bound together by a dynasty and had become a single constitutional state. Austria had not got this far, especially as the constitution drawn up for the German half of the empire in 1848–9 was first suspended and then abolished.

What advance there had been towards constitutionalism and political citizenship took place within individual states. This contributed towards deepening state loyalty and commitment. Such state loyalty would shape the way in which national loyalty developed, influencing people to oppose some and to favour other kinds of unification. Already in 1848 state parliaments and liberal ministries had reacted negatively against attempts by the GNA to impose 'national' policies upon them, even when in principle they approved of the formation of a national state. The state frontier demarcated a clear zone of legal and political citizenship by the 1850s but there was no obvious process taking this beyond state to national frontiers.

Social citizenship

Before the modern welfare state the only social right that mattered concerned access to poor relief. Poor relief was an affair of the locality – the parish in England, the *Gemeinde* in Prussia. The state might lay down common rules but the obligation to administer these rules fell upon the locality. Paupers should be returned to their locality for support. Given that for most people legal and political rights were remote and unimportant, this meant that citizenship that really mattered was attached to membership of the local community. That also meant that the most important boundary was the boundary of the local community. States did not need to police their frontiers to control the implementation of this right because they could depend upon local government. Except in a few growing towns, such enforcement did not require any separate or specialized policing as it was quite easy to distinguish strangers from natives. Only at times of economic crisis when there might be floods of beggars on the move would local authorities look to the state for some assistance in control and might there be established gendarme or military patrols to prevent the entry of or to deport foreign paupers.

The fact that this was the 'citizenship' right which mattered most to most people helps to explain the way in which citizenship as state membership was defined. For much of the nineteenth century the most frequently used term for state membership was that of *Heimatsrecht*, literally 'home right'. People acquired this right at the local level. It was acquired through descent (not birth) or permanent residence. In addition strangers could acquire *Heimatsrecht* through payment of a fee and satisfactory proof of economic independence.[14] All these features of 'citizenship' are unlike modern citizenship and can be explained in terms of the essentially local character of citizenship. Descent rather than birth mattered because it was quite likely that a child could be born outside the locality in which his or her father pursued his livelihood. This principle of descent had nothing to do with any ethnic principle or the later German principle of *ius sanguinis* as has sometimes been claimed.[15] First, descent carried no ethnic meaning; the principle not only excluded Germans of other states from *Heimatsrecht* but also Germans from other localities within the same state. Second, lawyers of the time displayed no awareness of such a 'law of the blood'. Third, the possibility of acquiring citizenship through permanent residence (and therefore of transferring membership from one commune to another) contradicted the principle of *ius sanguinis* which insists on the continuation of citizenship after emigration and into a new generation unless explicitly renounced. However, considered from the local perspective the removal of a person from one locality to another, and the taking up of a new occupation in the new locality, should involve a transfer of *Heimatsrecht*. Finally, given the functional character of such citizenship, it was essential that suitably qualified strangers could acquire it.

Of course, there was an important sense in which people were members of a territorial state as well as a locality. A German state could not, at least by the middle of the nineteenth century, deport its 'own' paupers. To have *Heimatsrecht* within a particular locality of the state was in effect to have state membership. However, that state membership was derived from *Heimatsrecht* and not the other way around. However, just as deportation agreements required some explicit definition of state membership (for example, in relation to those unfortunates who had no *Heimatsrecht*), so too did other changes after 1815 increase the importance of state membership and correspondingly reduced the importance of local *Heimatsrecht*.

First, reforming state governments could not accept that individual localities could impose different rules for the acquisition or loss of citizenship. Even if local autonomy was preserved, increasingly this was on uniform terms laid down by the centre. Second, increases in geographical mobility associated with economic change made it necessary to reduce the power of the locality to deny membership and, therefore, poor relief. In addition, some governments, notably that of Prussia, thought that economic growth required greater freedom of movement than was possible with existing communal constraints upon local citizenship. In 1842 the Prussian government passed two laws which made it easier for people to move from one locality to another and to qualify for poor relief in the new locality. On the same day as these laws came into operation, so did a third law which for the first time defined state membership (*Staatsangehörigkeit*). Once the local boundary had been eroded for the purpose of implementing social citizenship, it became necessary to make a clear distinction between citizens and foreigners at the level of the state frontier. The use of the term *Staatsangehörigkeit* made it clear that this was something more abstract and extensive than *Heimatsrecht* but it did not carry with it any connotation of active, political citizenship rights. A politically empowered citizen was a *Staatsbürger* rather than a *Staatsangehörige*.

The effect of these Prussian laws was again to increase the importance of the *state* frontier. German states treated each others' members as foreigners, making no distinction between them and non-Germans. A Bavarian was a foreigner (*Ausländer*) in Prussia and vice versa. Indeed, many migrant workers often found it easier to cross the border into Switzerland or Belgium or France than into another German state. Not surprisingly, such Germans when in exile did gather together as Germans from different states and this made them more aware of a German identity. However, they used that identity as a way of criticizing the 'real' Germany which was often depicted as a series of prison cells. To create freedom it was necessary not only to abolish the prison but also the walls dividing the cells from each other.

Nevertheless, the bilateral agreements on deportation and the multilateral agreement on freedom of movement and acquisition of citizenship did mean, by the 1860s, that even at the level of social citizenship which was the type of citizenship which concerned most people, there had developed a difference between the ways in which different German states related to another and how they related to non-German states. To this extent one can talk of a limited German 'political system' with some sketchily defined citizenship rights.

Frontiers and citizenship obligations

Taxation

Taxation also was a matter of the individual states seeking to create common systems throughout the territory. The construction of the *Zollverein* was accompanied by the abolition of toll barriers within individual states. The abolition of privilege was linked to a unifying of the fiscal system. The reduction of the power of local government over who received poor relief made state regulation of this matter increasingly important, often accompanied by some transfers between localities. Tax assessments and collection increasingly applied to individuals rather than collectivities such as families or communities, especially with the creation of a more geographically mobile labour force and individualized property rights. Finally, taxation was sometimes linked directly to citizenship rights, most notably with the Prussian three-class franchise. Once again all this would suggest that the state frontier acquired an increasing salience for many people, as fiscal citizenship was unified within the individual state but differed sharply from one state to another. However, this can only be put in the most general and speculative way as the history of citizenship from the fiscal perspective remains to be researched and written.[16]

Military service

France had improvised a citizen army during the Revolution and Napoleon had placed this on a systematic basis of conscription. With varying degrees of reluctance other states followed suit, first asking for volunteers and then moving towards universal conscription. Such an army was intended to replace a professional army made up of career officers (usually nobles) and soldiers recruited either as mercenaries and/or through the impressment of poor people, usually from the countryside. Prussia retained the principle of universal conscription after 1815 even though decline in army size meant that many exemptions were granted. Effective implementation of the principle was an important part of the reforms of the early 1860s, reforms which were crucial in bringing Bismarck to power (because of the constitutional crisis they provoked) and in winning the wars of 1864, 1866 and 1870–1 which unified Germany. The reforms were not popular, both because people did not wish to perform military service and because the expanded and modernized army remained under the control of the king rather than parliament. With the establishment of the North German Confederation in 1867 the Prussian model was extended to other states. Pressure to adopt this model was also applied to the southern German states which were outside the Confederation but had entered into a military alliance with Prussia. Such pressure increased anti-Prussian feeling in southern Germany up to the outbreak of war with France in 1871. Military citizenship, like fiscal citizenship, was imposed by the state on a reluctant population and was only extended to a national level through the power of Prussia.

The Bund *as a military alliance*

Prussian pressure after 1867 for common forms of conscription and training throughout the German states except Austria makes clear that its military system did not yet apply in other German states. Austria, the other military power within the *Bund* until 1866, maintained an army on more traditional principles of peasant rank-and-file and noble officers. The other German states, although required to maintain armed forces under the terms of the Federal Act, did not act zealously upon that requirement. These states looked to Austria and Prussia for protection from other major powers, and those two states had no interest in creating a multi-state army which might call into question the loyalty of their own soldiers.

However, it was the military factor which most closely associated a distinct function with a 'national' frontier. One of the intentions behind the creation of the *Bund* and a Federal Army was to prevent a power vacuum in the German lands which might tempt great power interference. In practice it was southern and western Germany, the region in which the splintered state system of the Holy Roman Empire had developed farthest, which was seen in this way. Indeed, for many contemporaries 'Germany' implicitly meant the smaller states, mainly to the south and west, apart from Austria and Prussia.

On the eastern frontier the Hohenzollern and Habsburg dynasties continued their traditional mission of defending Latin Christianity against Orthodox Christianity and Islam, though now in a more secular fashion, and they did so without any coordination or as 'German' powers. The Habsburg Empire maintained its military border zone against the Ottoman Empire. The frontier between Prussia and Russian Poland was the only walled frontier in the German lands, designed to be genuinely impassable. There was no 'national' content to these eastern boundaries. In the case of the Habsburg Empire the eastern boundaries with Russia and the Ottoman Empire simply divided different Slav groups from each other. Only for the Magyars of the eastern half of the Habsburg Empire might the state frontier be regarded as national, and even then only in terms of securing Magyar hegemony over a territory which included many non-Magyar speaking peoples. In the case of Prussia, there was a long tradition of ruling over Polish-speaking peoples in East Prussia, as well as the acquisition of further Polish subjects through the partitions of Poland – after 1815 reduced for Prussia to the province of West Prussia and the Grand Duchy of Posen. All this meant that the distinction between German and non-German in the east was not a sharp geographical one but rather a matter of culture, social rank and political rights. When there was an attempt to draw an ethnic frontier in Posen with disputes between Germans and Poles in 1848, this was a pragmatic solution to a local crisis rather than the expression of a settled view about national frontiers and how they should be demarcated.

On the northern and southern frontiers there was no serious threat. As in the east, so in the south the 'ethnic frontier' had little meaning as the Habsburg Empire extended well beyond zones of primarily German settlement. In the north the only contentious issue concerned Schleswig-Holstein, where both German and Danish nationalists rejected the idea of an ethnic frontier but disputed whether Schleswig should be wholly under Danish or wholly under German rule. In any case, Denmark did not represent a threat to the security of German states.

The only state which did represent such a threat was France. Not only was France a major power, even after defeat in 1815, but it was regarded as the most important threat to international and social stability (followed closely by fears of Polish risings in the Prussian, Russian and Austrian areas of occupation). Furthermore, France had pioneered the most definite view of the citizen/foreigner distinction and linked this to a sharp demarcation of the national frontier. It was in western Germany that Austria and Prussia co-operated most effectively to maintain certain garrisons and to co-ordinate military efforts. It was from memories and myths of the great 'War of Liberation' of 1813–15 that the most ringing affirmations of national frontiers were derived. 'Watch over the Rhine' became a tremendously popular song during the crisis between France and the German states in 1840.

Yet although this was the clearest example of a 'sacred' frontier, in many ways it had less functional significance than other boundaries between German states. With the signing of free trade agreements between France, Britain and the *Zollverein* in the early 1860s, for example, along with numerous agreements to maintain navigation along the length of the Rhine, there were fewer controls on movements of goods and people between France and Belgium and the west German states than there were between member states of the *Zollverein* and the Habsburg Empire or between Prussia and Russian Poland.

The Bund *as a political system*

In addition to the idea of a common military policy, the Federal Act also imposed the idea of common diplomacy. Member states of the *Bund* were not supposed to pursue independent foreign policies. Austria and Prussia sidestepped this provision by having territory located outside the *Bund* so they were both members and not members of the *Bund*. This option was not open to the other states (except when they were linked by personal union to a foreign ruler, as was the case of Hanover up to 1837 or of Holstein).

In external relations the German states, apart from Austria and Prussia, lacked diplomatic or military independence. In internal matters, Austria and Prussia could and did impose restrictions to the development of political freedoms. All this made politically active members of other states acutely aware that their states were dependent bodies within a broader national political system. Finally, as we have seen, in such matters as customs union (though this never included Austria), or mutual rights of emigration and citizenship, there did develop something of a national system. In all these ways, even without taking into account the growth of nationalist sentiments, one can see that the states of the *Bund* did form a loose political system rather than being a confederal association of equal and independent states. Nevertheless, for the everday lives of most Germans, the most important transformations in the period 1815–66 had been in the erosion of the local frontier and the increasing importance of the state frontier.

Frontiers and war

It is war or the threat of war which, under modern conditions, frequently give state frontiers a symbolic as well as a practical significance. The projection of Germany as a cultural nation that should control its own territory was most powerfully expressed in the war against Napoleon and during the crisis with France in 1840. Leaving aside the issue of how popular this nationalism was, one should note a number of things. First, war or fear of war across national frontiers before 1871 only really applied to the border with France and, to a much lesser extent, in the north with Denmark. There was no symbolic significance in the southern and eastern boundaries. Russia remained an ally of the Prussian state until 1890. The fact that Austria was a German power was a source of embarrassment to *kleindeutsch* nationalists because it meant that war in the south could not be couched in national terms.

Second, in the absence of war or diplomatic crisis, the most important frontiers for Germans were those of the individual states. On a day-to-day basis there was little that would make most Germans, especially poorer Germans, feel any national identity except probably in the negative sense of envisaging national unity as a means of freeing them from the restrictions of the individual state and those applied by the *Bund*. Better-off Germans, especially those who traded goods across state borders and who read national publications, might have a more positive view of a German state but they themselves usually exercised influence through individual states and would not be willing to give up that influence.

It is not surprising, therefore, that unification came about not through elite or popular political pressure but through war. In two of those wars, those against Denmark and France, there was a great deal of national enthusiasm. However, the crucial war was that of 1866 because it excluded Austria and made it clear that all national institutions from then on would be shaped above all by Prussia. That war was a cabinet war which dismayed elite nationalists and engaged popular loyalties only in terms of state membership or religion but not in national terms.[17]

However, these wars, especially that of 1870–1, altered the significance of state frontiers in two important respects. First, propaganda associated with the Franco-Prussian war led to a 'nationalization' of war aims. In particular the annexation of Alsace-Lorraine was associated not simply with argu- ments about national interests – strategic and economic – but also the elaboration of contrasting national myths. German nationalists claimed, in relation to Alsace, that the population was 'really' German by virtue of language and history. French nationalists claimed that the population was 'really' French by virtue of citizenship and loyalty. This led to the elaboration and counterposing of an 'objective' (ethnic, linguistic) 'German' view of nationality against a 'subjective' (citizenship, will) 'French' view. Ernest Renan's famous aphorism that nationality was a 'twenty-four hour a day plebiscite' was opposed to Heinrich von Treitschke's furious assertion that the Alsatians had 'forgotten' their true nationality. This could then be related to the distinction between the French citizenship principle of *ius soli* based on birth and the German citizenship principle of *ius sanguinis* based on descent.[18] Yet there is much to suggest that these distinctions were products of the particular crisis in Franco-German relations in the 1870s and 1880s. We have already seen that the principle of *ius sanguinis* was unknown to pre- 1871 German states. The French move away from *Heimatsrecht* to state citizenship was linked to radical changes of the Revolution. The clear enunciation of *ius soli* only came in the 1880s, in a state obsessed by relative population decline, thirsting to revenge 1871, and wishing to ensure the largest possible citizenship base for its military recruitment. The Franco-

Prussian war as arguably the first popular national war fought in Europe imparted a national meaning to the frontiers produced by German victory which had been missing before but it is typical of nationally informed histories that these differences were then essentialized and projected back into the past.[19]

Second, the conditions of warfare altered the significance of frontiers. The wars of post-1850 Europe – the Crimean War, the Franco-Austrian war of 1859, the German–Danish war of 1864, the Austro-Prussian war of 1866 and, most dramatically, the Franco-Prussian war of 1870–1 – were underpinned by the new capacities of rail transportation and the mass production of modern weapons. Military mobilization and movement before 1850 had been a much more leisurely and seasonally conditioned process. Now the state frontier was the place on which military preparations focused. On one's own side of the frontier the appropriate technical and industrial preparations were made. Then movement across the frontier was not only the trigger for war but, if carried through quickly and massively enough, effectively determined the outcome not just of a battle but the whole war. Military preparation behind the frontier and intensive surveillance of the frontier therefore greatly increased in importance.

Germany as a nation-state

Introductory remarks

The German nation-state of 1870–1 had been created suddenly, by force of arms. Amongst certain elites there had been a longer tradition of national awareness. At a popular level a certain amount of inter-state movement and shared rights across state frontiers within the German 'political system' had created some sense of national identity which had been further stimulated by war against foreigners, in particular the French. All this could give to the new state and its frontiers, especially those on the west, a widespread national meaning. Furthermore, the new state-formation had been accompanied by a national and constitutional process. The states annexed to Prussia in 1867 were not incorporated in the manner in which Frederick the Great had incorporated Silesia over a century earlier by monarchical assertion plus tacit recognition of existing customs.[20] Rather it was incorporated into Prussian constitutional arrangements. In addition the North German Confederation provided a constitutional expression of the national idea designed to replace the German Confederation, to integrate non-Prussian territories and to exert an attraction upon the south German states remaining outside this German framework. The constitution of the Second Empire, agreed in January 1871, completed that process at the formal, constitutional level. Thus an expansion of citizenship rights was an integral part of frontier revision in a national direction.

On the other hand, unification had involved the expulsion of one of the two major German powers, Austria, from the new state. It had been achieved by an authoritarian, Protestant Prussian state which aroused suspicions and antipathy amongst democrats, Catholics and non-Prussians. Finally, for most Germans, their everyday life was more shaped by functions performed at state or sub-state level than at a national level. A lot needed to change for national frontiers to acquire a central meaning for most Germans.

The acquisition of citizenship

The constitution of 1871 has a preamble which reads like a treaty between the states making up the Second Empire. Imperial powers over war, peace, tariffs, etc. were asserted but the general rule was that, unless otherwise specified, powers belonged to the individual states. German citizenship was a derivative of state citizenship. Germans did not have common rights throughout Germany; rather a Saxon could enjoy the rights of a Prussian when in Prussia and vice versa. The individual states controlled naturalization procedures which they often applied in different ways. The only genuinely national citizenship right was political: the right of all males above the age of 24 to vote in Reichstag elections.

The adoption of *ius sanguinis* as the German principle for the acquisition of citizenship only came in a law on *Staatsangehörigkeit* in 1913. In part the law was codifying the emphasis on descent rather than birth which had been a feature of traditional *Heimatsrecht*. However, it forced that emphasis in a particular direction. First, it overrode state autonomy on naturalization by making it possible for any one state (through its minister of interior) to block naturalization in any other state. This was motivated by the concern to shut out Jewish and Polish immigrants from German citizenship. Second, it stipulated that Germans would remain German after emigration unless they explicitly renounced German citizenship. The general regulation, which many liberal and socialist deputies had demanded in order to unify citizenship procedures throughout the German states and to take account of the settlement of Germans in overseas colonies, had been hijacked by a vociferous nationalist minority supported by the government. The law was important, not least because it underpins the German principle of citizenship up to the present. Nevertheless, it was not the product of some long and consensual 'German' view of citizenship and nationality but rather a contingent if far-reaching product of political conflicts and currents generated in Wilhelmine Germany in the decade or so preceding the outbreak of the First World War.

The erosion of state frontiers

The law of 1913 sought to provide a national framework for the acquisition of German citizenship rather than leaving this as a matter for individual states. However, except for the political right of voting in Reichstag elections and the obligation to serve in the army, legal, political and social citizenship rights varied from state to state. More fundamental than these limited common rights and obligations for fostering a common sense of national identity were the processes which eroded the importance of state frontiers in the lives of many Germans and conversely increased the importance of the national state and its frontiers.

The regime, as part of a project of integrating industrial workers into the existing state without too many political concessions, set about providing social benefits.[21] The provision of welfare rights such as old age pensions and sickness and injury benefits made the imperial state much more important for many Germans. The central state set up numerous agencies, for example concerned with public health, which overrode state government where necessary and helped alter and raise expectations about the quality of life amongst many Germans. Legal equality was largely achieved and taken for granted and by the end of the century the civil law had been codified at a national level. Political rights were not extended at a national level and, indeed, were reduced in some states through measures restricting the franchise. Nevertheless, these national political rights came to mean much more. The proportion of the electorate voting rose from 50 per cent to 85 per cent between 1871 and 1912 which, coupled with population increase, tripled the number of those voting. Furthermore, the Reichstag became much more significant along with the growth in the powers of the imperial state, much of which required legislation. Political parties had shifted from a regional or state base to being genuinely national organizations with programmes and firm discipline. The growth in imperial tasks exposed the inadequacy of imperial financial arrangements and that led to crisis and an extension in the fiscal powers of the imperial state. In common with many other modern states the tax base shifted increasingly to direct taxes on income and property which in turn made individuals more aware of national authority. All this meant that in internal and functional terms individual state frontiers came to mean less and experiences of individuals between one part of the national state and another became more uniform.

Such experiences became more communicable across state frontiers. Compulsory education and high literacy rates provided the basis for the production of popular literature, be it school textbooks or mass circulation periodicals. The growing and increasingly youthful population was increasingly on the move. People migrated longer distances than before yet at the same time the growing prosperity of Germany meant that less of this took the form of emigration and more of internal migration. In some places, especially Berlin and the boom towns of the Ruhr area, maybe as many people as half the total population passed through in the course of a year.[22] Extensive internal migration and reduced emigration meant that state frontiers meant less and the national frontier came to mean more.

The growing importance of the national frontier

The construction of a mass national society and public opinion increasingly concerned with the activity of the imperial state created a receptivity to arguments about the importance of the frontiers dividing Germany from other territories. This receptivity has to be placed within the context of the growing functional importance of national frontiers.

The *Zollverein* had been a national institution of great economic importance which prefigured Prussian domination and Austrian exclusion before the decisive war of 1866. However, it had been regarded principally as a vehicle of economic liberalism, especially in the 1860s and the early 1870s. The slump of 1873–4 and the period of depression which followed led to protectionism, inaugurated in 1879, increased in the mid-1880s, temporarily relaxed in the 1890s but then renewed at even higher levels from 1902. Protectionist arguments used the language of economic nationalism, depicting the territory of the nation-state as an economic unit with a distinct set of interests that the state should foster. The world beyond the national territory was portrayed as a threat to the national economic interest. Protectionism has a spiral logic; if interests are damaged by the protection afforded to others but do not feel able to reverse the original measure of protection, then they demand protection for themselves. Economic interests in other countries are stimulated to demand similar responses from their states.

Thus protectionism tends to become a self-fulfilling prophecy. The competitive model of classical political economy is projected from the level of individuals to that of nations. This helped establish nationalist ideologies which used the language of struggle, often adopting Darwinian images and sometimes suggesting a biological, racial basis to such struggle. At the functional level protectionism required an increase in frontier surveillance and policing and habituated people to experiencing the frontier as a definite moment of control, a sharp line to be crossed.

The same applies to population movement. Here one has to distinguish two different ways in which the frontier is operated. In some cases, especially where there were fears of collective incursions – be it armies or bands of criminals or beggars – the emphasis was upon physical control at the frontier or frontier zone. The appropriate enforcers were police and military patrols.

This kind of frontier control has a long history which predates the emergence of the frontier as a sharply defined line.

More modern is what might be called bureaucratic control. One can see this developing in conjunction with more elaborate definitions of state membership developed in the bilateral agreements on deportation. Control became a matter of identifying who was and who was not a member of the state. The frontier mattered because the state was identified as a territorial entity and one acquired membership of the state through birth or descent or residence within that territory. To prove membership or temporary entitlement to be within the state territory required documentation. Foreigners coming into the territory of the state were not subject to physical control at the frontier but instead required to present themselves to the first police station or some other designated place to have their papers checked. The importance of local as well as territorial state membership in the early nineteenth century meant a similar requirement was placed upon internal migrants. With the erosion of state frontiers but the enhanced importance of the national frontier in Wilhelmine Germany this bureaucratic surveillance came to focus upon non-nationals.[23]

Bureaucratic surveillance increases with the level of movement across the frontier. Increased cross-frontier movement intensifies the making of national/non-national distinctions. The German eastern frontier in the last decades before 1914 provides a good example. The construction of elaborate social citizenship rights, along with national political rights and legal equality, all within the context of an expanding economy, made German citizenship a valuable possession. At the same time, wage-level differences stimulated a mass internal emigration from the farms of eastern Germany to the mines and factories of western Germany. East German farmers found themselves confronted with an acute labour shortage, especially at the most demanding times of the season. The answer was provided by the seasonal influx of Polish-speaking labourers from Russian Poland. The German state, local landowners and nationalists wished to ensure that these labourers did not come to reside permanently in Germany or manage to acquire German citizenship. This was partly because of the costs and benefits of such citizenship. It was also related to the way in which the formation of the German nation-state had turned the Polish-speaking subjects of Prussia into a national minority. It was enough that these Polish speakers possessed German citizenship (thus being able to vote for nationalist deputies to the Reichstag) and were perceived as a major threat to German control in the east without adding to their numbers. The citizenship law of 1913 with its principle of *ius sanguinis* was intended in part to shut out such Polish immigrant workers. However, in this case the bureaucratic controls were harshly and actively enforced as well as being supplemented by aggressive policing of the frontier. In addition there was the economic policing of the frontier after the Russo-German tariff agreement of 1894 had lapsed and the German government enforced high protective tariffs on Russian agricultural products.

The final function emphasizing the significance of the eastern frontier was military. The expansion of the army up to the mid-1890s and then again in 1913, along with the construction of a battleship navy from 1898, affected millions of Germans. Adult males underwent the experience of military service. Whether that increased a positive sense of national identity and loyalty is a moot point but it certainly made people more aware of the national and military dimension in their lives. Given the expense of military provision and the importance of public opinion and the Reichstag to securing the appropriate legislation, the government and various interests engaged in active propaganda to secure mass support for military expansion and modernization. Germany was, rather contradictorily, portrayed both as a country under threat from aggressive neighbours and as an incomplete world power which needed to build up its strength abroad by such policies as the acquisition of an overseas empire. On the eastern frontier the failure to renew the Reinsurance Treaty with Russia in 1890 meant the lapsing of a friendly Prussia/Germany-Russia relationship which had operated continuously since 1813. Within four years Russia had concluded an alliance with Germany's most inveterate enemy, France. Whereas Russia had previously figured as a threat only in the rhetoric of liberal and radical opponents of the German state, official nationalism now came to emphasize the Russian threat, in part taking over the images – used by liberals and radicals – of Germany as advanced, modern and civilized and Russia as backward and barbaric. It was an imagery which would be used to good effect by the German government in the July crisis of 1914 to help persuade the anti-war socialists – the largest party in the Reichstag – to support the war effort. Germany as a threatened national territory was an image with more appeal in 1914 than that of Germany as a power which must expand in order to survive. The frontier as a line of defence was the principal message, even if in reality the frontier was employed as the jumping-off point for a war of offence.

German frontiers in the twentieth century

German frontiers meant little during the First World War. The war was fought on the territory of Germany's enemies. The elaboration of war aims which envisaged the creation of an overseas empire and the extension of German power in Europe either directly or through the creation of satellite states made it clear how superficial was any territorial conception of

Germany, at least for those who ruled Germany. The kind of settlement Germany would have imposed had it won the war can be inferred from the savage treaty of Brest-Litovsk concluded with Russia in 1918.

Of course, Germany lost the war. War had been a great nationalizing experience as economic and military mobilization, official political consensus and unremitting demonization of Germany's enemies shaped the everyday life of all Germans. From 1916 there was increasing discontent with the war and the policy of the government, and the argument that the war should really be about nothing more than the defence of existing national frontiers began increasingly to be heard. Nevertheless, the 'peace without annexations' argument was as resolutely framed in national terms as any expansionist programme.

Defeat placed power in the hands of the peace party, a coalition of socialist, Catholic centre and left-liberal politicians. Accordingly they presented the case for preserving the integrity of the 1914 German frontiers to the victorious allies. They argued in vain. The French were determined to take back Alsace-Lorraine. The Danish minority insisted on the transfer of Schleswig. Polish nationalists reversed the partitions of the eighteenth century.

Yet in certain respects the Allies *did* accept the idea of a national territory. Although other parts of Germany were taken from her, such as the Saarland, or temporarily occupied, as in the Ruhr, or subject to international control, as with the demilitarized Rhineland, there was a general acceptance of the idea that Germany as a national state should continue in existence, along-side other nation-states. It was all right to break up the Ottoman and Habsburg empires because the national argument supported that strategy. The same argument, however, ruled out any partition of Germany. Admittedly, the Allies did not take the logic of nationality and plebiscites to its logical conclusion for Germans because that would have led to the attachment of the rump state of Austria to Germany.

The loss of territory in the east, north and west rendered Germany ethnically more homogeneous. This was enhanced by the immigration of German minorities from the lost territories. Taken together, the 1913 law based on *ius sanguinis*, the experience of war and then of defeat and territorial loss all intensified the sense of German nationality as ethnic. There was also a widespread consensus amongst Germans that the settlement of 1919 was unjust. By 1933 the idea of Germany as an ethnic nation-state which had to assert its autonomy in central Europe was a matter of broad political consensus. That is not to argue that the specifically racial nationalism of Hitler and the national socialists commanded popular support, but without that underlying consensus it is difficult to explain Hitler's rise to power and his great popularity until the war turned against Germany.

Yet there was a constant tension between defensive and offensive nationalism. For the advocates of defensive war in 1914, a 'just' peace settlement in 1918–19 and, failing that, for appropriate revisions to that settlement, the national territory was regarded as that of 1871, perhaps without border areas in which the majority of inhabitants had another national affiliation. Furthermore, the founders of the Weimar Republic created an institutionally more integrated national state than had existed before. Citizenship was membership of the national state, not of a member state. Political and social citizenship rights were greatly expanded, both nationally and within the states which were required to have democratic constitutions. The principal external demands were less for frontier revision than for providing Germany with the same rights as other states – for example by removing reparation burdens and controls over military policy and enabling areas like the Saarland to opt for rejoining the German state.

These were demands which more aggressive nationalists like Hitler could pursue and, in the early period of the Third Reich, his policies could be seen as highly successful within this framework. This helped secure his popularity at home (especially as revisions were achieved without war). Arguments for the justice of such revisions also formed the moral basis of the appeasement policy of the British government.

However, the nationalist ideology of Hitler was fundamentally opposed to the idea of national frontiers. It did not make expansionist claims on behalf of the territorial nation-state but instead in the name of the racial nation. Objectives of *Lebensraum*, racial expansion and racial empire destroyed the meaning of frontiers, national or otherwise. Frontiers would be swept aside along with the other institutional features of the modern nation-state. The barbarism that accompanied this expansionism discredited any idea of a German nation-state. Whereas the national argument had made German partition unthinkable in 1919, there was little protest against the de facto partition of Germany after 1945. German frontiers came to take on a new meaning after 1945. The idea of Germany as an ethnic nation based upon the *kleindeutsch* solution of 1871 predominated. Austria was able to pursue a separate path as if it had not been an integral part of pre-1871 Germany and of a greater Germany between 1938 and 1945.

The continuing commitment to the national idea was expressed in the doctrine of 'two states, one nation' to which both the Federal Republic of Germany [FRG] and the German Democratic Republic [GDR] subscribed.[24] In the case of the FRG the commitment to the national character of the earlier German state was made clear in the provision in its constitution, the Basic Law of 1949, which defined German citizenship (meaning citizenship within the FRG) by descent according to the 1913 law and projected this definition onto the territory of Germany as of 31 December 1937, that is, excluding

Austria but including some of post-1945 Poland as well as all of the GDR. The Basic Law also offered citizenship to the *Volksdeutsche*, ethnic Germans who continued to live in other parts of eastern Europe. This was linked to the assertion in the Preamble of the Basic Law that the people who founded the FRG were acting ' . . . on behalf of those Germans to whom participation was denied' and that they called upon: 'The entire German people . . . to achieve in free self-determination the unity and freedom of Germany.'[25]

Taken out of context and considered unsympathetically this could be read as the expression of an expansionist nationalism in which the state territory of the FRG was treated as the jumping-off point for the restoration of a united Germany. That would be a gross misrepresentation. The Basic Law could only have been adopted with the approval of the Western powers which were resolutely opposed to any such expansionist programme. The reference to 'free self-determination' made it clear that German unity was predicated on the basis of liberal democracy. The extensive definition of citizenship had a personal, not a territorial significance. It meant that the FRG offered itself as a place of refuge for Germans placed under the undemocratic rule of the GDR and other East European states.

Indeed the 'western' orientation of the FRG killed off any realistic pursuit of national unification after 1949. For Adenauer, good relations with France and other Western European powers and, above all, the maintenance of ties with the USA, meant that the FRG had to commit itself wholeheartedly to the Western alliance. State frontiers with Western powers were diminished in importance as a matter of policy, embodied in such measures as the formation of the Iron and Steel Community and later the Common Market. At the same time, this Western commitment during the period of the Cold War intensified the importance of the frontier between East and West. This frontier acquired a notorious and fixed character in 1961 when the GDR, concerned to block the loss of valuable labour power to the west where emigrants could immediately acquire FRG citizenship rights, built the Berlin Wall. The rigorously patrolled frontier between the FRG and the GDR divided a nation and was part of a broader supranational barrier between communist and capitalist states, each dominated by a world power which in turn reduced the significance of state frontiers within each of the power blocs. Within those two blocs a new generation of Germans grew up in the two German states with very different experiences and concerns.

Yet nationality and frontier continued to combine in important ways. The FRG citizenship principle of *ius sanguinis*, while expansive in relation to East European Germans, was restrictive vis-à-vis immigrant workers from southern Europe who streamed into the prosperous German economy from the 1960s, increasingly settled permanently, and established families. The ethnic definition of citizenship and nationality and the denial of citizenship to long-term residents appeared increasingly illiberal, not least to many Germans.

The expansive ethnic aspect of citizenship was a constant temptation to East Germans to seek escape, even when this was perilous after 1961. The frontier was a physical barrier to east–west movement but it did not prevent West Germans visiting friends and family in the GDR, especially when the GDR government could exploit this to earn hard currency. Equally it did not prevent all manner of loans and economic subventions from west to east.

This intensified with the policy of *Ostpolitik* initiated in the mid-1960s and especially associated with Willy Brandt. The policy accepted the permanent division of Germany precisely in order to increase positive contacts across that frontier. This vividly expresses the ambivalent nature of frontiers. Those right-wing nationalists who most vehemently denounced the division in the name of national unity did so in a way which blocked any chance of national reconciliation and co-operation. This makes abundantly clear how modern frontiers are first and foremost attributes, functional and symbolic, of sovereign states and only secondarily national boundaries.

There was an asymmetry in the 'German–German' relationship developed under *Ostpolitik*. The FRG pursued the policy for political reasons: to reduce tensions and also the dangerous dependency of the state upon US policy. The GDR pursued the policy for economic reasons: to obtain resources for its failing economy. The policy reflected the increasing confidence and independence of the FRG, the insecurity and weakness of the GDR regime. The contrast was clear to ordinary Germans on both sides of the frontier, through personal visits from west to east as well as the exposure of East Germans to the television programmes and consumer culture of the West.[26] It deprived the Berlin Wall of any legitimacy and made it wholly dependent upon the policing power of the GDR state. That policing control was drastically weakened during 1989 as other Eastern European states dismantled some of their frontier controls, enabling GDR citizens to emigrate to the FRG via Hungary and Austria. When control over the Berlin Wall finally broke down in November 1989 it led rapidly to a general crisis of the GDR state which could only be resolved by a political takeover by the FRG.

One remaining national frontier problem concerned the frontier of the newly expanded FRG with Poland. The Basic Law of 1949 had taken the territorial norm of Germany as that of December 1937. In the absence of any formal peace settlement after 1945, the Oder-Neisse line of division between the GDR and Poland had never been accorded international recognition. The pressure of German immigrants fleeing from those parts of post-1945 Poland on domestic FRG politics, especially within the governing party of the CDU, had made it impossible for conservative German politicians to recognize the Oder-Neisse line. Yet by 1989–90, although Kohl held out against acceptance for some time, this frontier did not have the negative symbolic significance of the Berlin Wall. International pressures were such that the FRG government quickly recognized the existing frontier between Germany and Poland.

For post-1989 Germany there have developed two state frontiers with different meanings. There are the frontiers between the Member States of the European Union. German policy is to diminish the significance of these frontiers. There are the frontiers with states which do not belong to the European Union. German policy is to increase the significance of these frontiers, which have been drastically weakened as barriers with the collapse of communism, by tightening up on its generous asylum laws and in other ways. I will not pursue these issues as they are considered in far more detail by other contributors to this book.

There is one other frontier which needs mentioning in relation to Germany and that is the internal frontier of the *Länder*. The earlier sketch of nineteenth-century Germany demonstrated how the state frontier increased in importance between 1815 and 1866 as local frontiers were eroded, and that although this process was also associated with some increased meaning in national frontiers (those of the *Zollverein* and of the *Bund*), national frontiers did not mean a great deal by the time the German nation-state was formed. I have also argued that the state frontier continued to mean a good deal under the Second Empire as many functions, including that of defining and acquiring citizenship, were state-based. State autonomy continued to be of importance in the Weimar Republic even if uniformly regulated by a more powerful and integrated national authority. Such institutional autonomy was destroyed under the Third Reich, less through any process of institutional centralization than through the destruction of all institutionalized forms of rule.

The FRG deliberately renewed the federalist tradition. One of the major flaws in the early system of federalism, the overwhelming size and influence of the Prussian state, was removed through the loss of most of Prussian territory to the GDR and Poland. One advantage of the federalist system of *Länder* was that it could be used to stress the incomplete and provisional nature of the FRG. Indeed the Preamble to the Basic Law declared it to be the creation of the inhabitants of the *Länder*. One way in which other parts of Germany could join the FRG was to attach themselves as additional *Länder*. That was indeed to be the legal mechanism by which the FRG and the GDR were brought together in 1990.[27] It entailed the artificial creation of *Länder* in the former GDR, and there are problems about the viability of some of these units. However, in the case of the original FRG the federalist system means that internal frontiers really mattered, for example in education.

This is significant also for the way in which it shapes German understanding of European integration. The federalist tradition, along with the policy of transferring various functions to supra-state institutions such as the European Union, mean that integration is not regarded as a zero-sum game in which indivisible sovereign powers of the state are removed to some other level and institution. This is very different from the common British, or rather English, understanding of state power in which internal frontiers and units of government have been comparatively unimportant; where frontier policing rather than bureaucratic surveillance has been the principal method of dealing with immigration, and where a long history of territorial integrity as well as great power status has firmly entrenched the idea of the nation-state with a sacred territory and an indivisible sovereignty. The utterly contradictory meanings of the term 'federalism' within German and English political rhetoric makes vividly clear the gulf between these political cultures.

Germany and Europe

Germany in relation to Europe

Every phase in the way in which 'Germany' has been organized relates intimately to a broader European pattern. The eighteenth-century Holy Roman Empire was a product of a general peace settlement of the mid-seventeenth century and was in part guaranteed through the membership of non-German princes who ruled a part of the imperial territory through personal union. The Holy Roman Empire was destroyed when French power undermined the European equilibrium which sustained it. The next 'German' institution, the Confederation of the Rhine, was a Napoleonic creation, set up for financial and military purposes. The succeeding German institution, the *Deutsche Bund*, was in part modelled on the Holy Roman Empire and the Confederation of the Rhine and was the product of a general European settlement. The *Bund* ceased to be viable as that settlement broke down with the outbreak of the Crimean War. The wars of unification directly involved the non-German states of Denmark, France and Italy and was passively supported or at least tolerated by Britain and Russia. Only the second phase of the war of 1870–1 started to turn some of that opinion against Germany. Only in 1894 did Russia enter into an anti-German alliance and in Britain strong anti-German sentiments were only generated through the sense of threat posed by the navy-building programme at the turn of the century.

By 1910 the idea of Germany as a threat to Europe had quickly become widespread and was made manifest through two world wars. The response took the form of anti-German alliances – the other major powers before 1914, an alliance of the new nation-states with France and Britain in the inter-war period, and a global alliance by 1941–2. Gradually such alliances shifted from being pragmatic affairs of national states to ideological blocs in which the national interest was subordinated to a broader value system such as that of liberal democracy or communism (themselves allied as anti-fascism). These broader alliances, dividing into the Cold War division between communism and capitalism, shaped the way in which Germany was organized after 1945. Only when that pattern broke down with the crisis in the USSR under Gorbachev, a crisis which rapidly undermined the power of the satellite communist states, could yet another way of organizing German territory be attempted.

Every phase has involved changes in both the frontiers and the institutional organization of Germany. Unlike those states which became national through nationalizing the territory and institutions of the existing state, such as France, Spain and Britain, there has never been a fixed political territorial meaning to Germany. The territorial definition of Germany in each phase was not so much the product of nationalist programmes as of state power: France between 1803 and 1812, the assembled powers of Europe in 1814–15, the Prussian capacity to exclude (but not conquer) Austria in 1866. Only after 1871 did Germany acquire a fixed territorial meaning and this was a combined result of political unification, the growth of a mass national society and increasing tensions with other national states. However, these frontiers were ones that the regime itself, and the most aggressive nationalists, wished to transcend in 1914 and 1939 and it led, through defeat, to truncation and partition respectively. This raises acutely the question of why most Germans should ever have considered any particular frontiers 'national', something I will consider in the final section of this chapter.

Germany as typical or unique

In the period up to 1914 Germany shared certain features with nation-state formation elsewhere. The most obvious similarity is with Italian unification which was also closely linked to the wars of German unification in 1866 and 1871. Yet there were important differences. Piedmont depended heavily upon other powers to achieve dominance, she was much weaker in relation to the rest of Italy than Prussia was to the rest of Germany, and there was much greater economic, social and political fragmentation in Italy. Also comparable is the USA where the Civil War led to a degree of political and economic unification of the various states on federalist lines that had been inconceivable before.[28] But that was a civil war in which national ideas were used very differently from the German case and which did not involve planting one powerful nation-state down in the middle of a multi-state system with all the security problems involved.

Germany differed clearly from other types of nation-state formations.

States like Spain, Britain and France were formed more through the process of internal consolidation or colonialism over a longer period of time and often associated with the development of overseas empires as well. Nevertheless, under nineteenth-century conditions what these cases shared in common with Germany, Italy and the USA was the assumption that nation-state formation involved processes of territorial integration, even if not unification, and was associated with ideas of progress through the creation of a liberal economic and political order. One of the reasons German unification initially was welcomed by liberals in other countries was because of its apparently progressive qualities.

The aftermath of the First World War undermined such assumptions.[29] New national frontiers were the product of an imposed peace settlement which multiplied the number of states and thereby reduced their size. Instability now was not just a matter of the threats large states posed to one another but also of the multitude of conflicts that could envelop a host of smaller states, each with troublesome national minorities. Boundary problems now were not just a matter of regulating movements of goods and peoples across state frontiers and constructing effective military security, but also involved contention between states about national frontiers, especially when one state took up the cudgels on behalf of 'its' national minority in another state.

Whether this multitude of nation-states could have produced a stable political system, including the regulation of frontiers and population movements, we will never know. The simultaneous recovery of the USSR and of Germany, along with the inability of the Western powers to control the consequences of such recovery, led to the destruction of the autonomy of the new states of Versailles, though not before (with the exception of Czechoslovakia) they had all removed their originally liberal democratic systems of government.

The result was a series of supranational blocs dominated by one major power: a short-lived German imperium, the Soviet system with its satellite states and the Western alliance. In the first two a central dictatorial state drastically reduced the meaning of national frontiers. In Western Europe, encouraged by the USA but also related to a widespread feeling that frontiers were fragile and unstable and that war led to mutual misery, the attempt was made to construct various supranational institutions which diminished the significance of national frontiers and sovereignties. In Eastern Europe the collapse of the USSR has led to the proliferation of many successor states which define themselves, more or less successfully, in national terms. All these post-1918 developments involved either nation-state formation through separation or the construction of supranational institutions, political processes unlike that leading to a German nation-state.

One comparison that might prove instructive is between the multi-state 'political system' of pre-1866 Germany and the European Union. Germany then, like the European Union today, consisted of individual states which commanded a good deal of loyalty and whose frontiers had real significance. Those states sought to construct some kinds of common citizenship, although only on the basis of what had been achieved at state level, and most of them had entered into a customs union. The ideas of citizenship which developed were not invested with the sacral and unique qualities attached to nationality within a sovereign nation-state but were seen as ways of con-structing and regulating various rights and obligations. These rights and obligations were not regarded as an indivisible bundle secured under one sovereign state but as distinct from one another and therefore amenable to different solutions at different levels and within different boundaries. There were, of course, important differences as well. The pre-1866 German states, even if able to generate some kind of common identity and loyalty amongst their subjects, clearly did not have the popular and communal character of modern European nation-states. Also there is no single, or even pair of dominant states within the European Union to compare with Prussia and Austria within the *Bund*.

It is also worth noting how a sense of national identity developed within the new German state. I would argue that German frontiers only acquired 'sacral' qualities not because these had any enduring place in a popular and well-established national tradition but because the development of a modern national society after 1871, the convergence of important functions upon the linear state frontier, and tensions with other nation-states combined to generate a sense of threat to national territory. That is why the western frontier was earliest and most sharply defined, why the northern and southern frontiers never acquired comparable symbolic significance, and why the eastern frontier acquired this only in relation to protectionism, population control and military security, replacing rather a sense of a mixed ethnic zone in which German culture should predominate.

This points to a tension between the functional ways in which the modern state has come to concentrate many of its controls upon the state frontier and the symbolic way in which ideas of sovereignty and nationality have invested the idea of a national territory with a sacred quality. Frontiers were vehemently defended less because the particular lines on the map had symbolic significance for all but the local population and more because they came to stand for state sovereignty which was equated with national independence. It is no coincidence that the idea of clear national frontiers had the least significance during the most aggressive phase of German nationalism, under the Third Reich, and that this was also a regime which undermined the idea of the state as a clearly defined and sovereign institution. What this suggests is that if one takes away the identification of the sovereign state with nationality and treats European frontiers as one functional boundary amongst others, this might both undermine the insistence on a 'sacred' national territory but also prevent such a sentiment being translated from the national to the European level.

Nation-states were not predestined, certainly in the particular forms they have taken, and the German case shows that they can be unmade and remade.[30] The idea of the nation-state as the only political community of significance is recent and incompletely realized. If citizenship can be stripped of the sacral and unique qualities with which it has been invested by the nation-state, where citizenship, nationality and territorial sovereignty have been fused together, it can open itself up to pragmatic and pluralist operations. This would have implications for frontiers amongst other practices. Just as railway journeys through small German states increased a sense of their obsolescence, so can increased and freer movement across nation-state boundaries, facilitated by and mediated through a severing of the citizenship/nationality nexus, have a similar effect in Europe today. Modern German history suggests that national frontiers which have been associated with crucial practical as well as overwhelming symbolic significance, and treated as strong and enduring fixtures, are actually contingent and changeable, both on the ground and in the minds of people.

Notes

1. For a survey of such approaches, see Anderson (1996), especially pp. 34–6 on the issue of 'territorial ideologies'.
2. In Breuilly (1993a) I classify these three different nationalist strategies as reform, unification and separatist. See especially pp. 9–14.
3. For introductory surveys in English, see Hughes (1988) and Schulze (1991).
4. I develop this argument in greater detail in Breuilly (1996) and (1997).
5. Marshall (1950). For a recent collection of critiques of Marshall's work, see Bulmer and Rees (1996).
6. I consider this in greater detail in Breuilly (1992a).
7. This assumes that the North German Confederation of 1867 was a single state, outside of which existed Hesse-Darmstadt, Baden, Bavaria, Württemberg and the German part of the Habsburg Empire. After 1871 there existed only the German Second Empire and the German part of the Habsburg Empire.
8. Though the changes were real and radical one should not overstate them. In some cases certain kinds of privileges either continued, or were invented or reinvented, such as those associated with imperial land grants made by Napoleon in the new Kingdom of Westphalia. Some satellite rulers, such as Dahlberg, Grand Duke of Frankfurt, managed to preserve traditional frontiers for sub-state units of government, even if adopting the new nomenclature of departments and

cantons. For a good English language survey, see Sheehan (1989), chapters 4 and 5.

9. See Demandt (1990), especially the introduction by the editor.

10. For this and much else that follows, see Fahrmeir (1997). Where Fahrmeir's interpretation challenges arguments I advanced in Breuilly (1993b), for example on the question of freedom of movement under the terms of the Federal Act, I would not accept Fahrmeir's views.

11. Given the sensitive nature of these issues the constitution did defer some of them for later legislation but there is little doubt that the majority of the parliament assumed that if German citizenship was to mean anything it required that much of the legal autonomy which existed at state and sub-state level be swept away in the name of national legal equality and unity.

12. I follow the account in Fahrmeir (1997).

13. See the studies in Rumpler (1990).

14. This is a very brief and sweeping summary of a complex subject and is indebted to the detailed treatment by Fahrmeir (1997).

15. Fahrmeir (1997) argues this point persuasively against interpretations such as that offered by Brubacker (1992).

16. For an exceptional and pioneering sketch of some of this subject, see Wehler (1987), pp. 369–80.

17. I develop these points in greater detail in Breuilly (1996).

18. The argument is developed by Brubacker (1992).

19. The most recent exploration of these counterposed national stereotypes and hatreds, one which argues for a continuity between the revolutionary wars and the First World War, is Jeismann (1992).

20. For pre-modern forms of frontier revision through annexation, see Greengrass (1991).

21. See Mann (1996).

22. This is not to say that half of the people stayed less than a year. If only one-twelfth of the population stayed in a town for an average of two months while the rest remained in place for a full year, this could suggest a 50 per cent turnover. Furthermore, it may be that the same people came and went a number of times over the course of a year. The statistics are themselves problematic and, even if good, the fact that they do not plot the movement of individuals but only of total numbers makes it very difficult to know just what the patterns and extent of geographical mobility were.

23. One might speculate on an inverse relationship between bureaucratic and physical control. The twentieth-century practice on the British mainland of close physical controls coexists with the lack of any requirement for people to carry identity cards; such a requirement in much of continental Europe (along with a further bureaucratic layer of registration requirements at hotels and other places of residence) coexists with less thoroughly patrolled frontiers.

24. This doctrine was finally abandoned by the GDR in its constitution of 1974.

25. Quoted from Hucko (1987), p. 193.

26. This exposure was as much, if not more, to US culture as to that specifically of West German culture, as a very large part of West German television fare consisted of dubbed American programmes.

27. Article 23. See Hucko (1987), p. 202.
28. For a recent comparison between these two cases, see Degler (1992).
29. A central argument in Hobsbawm (1991) and (1994).
30. How this affects the way in which we should relate the national idea to the long-term history of a modern nation is considered by a number of historians in Breuilly (1992b).

Bibliography

Anderson, M. (1996) *Frontiers: Territory and State Formation in the Modern World*, Cambridge: Polity Press.

Breuilly, J. (1992a) 'Conclusion: National Peculiarities?', in J. Breuilly (ed.) *Labour and Liberalism in Nineteenth-Century Europe: Essays in Comparative History*, Manchester: Manchester University Press, pp. 273–95.

Breuilly, J. (1992b) *The State of Germany: The National Idea in the Making, Unmaking and Remaking of a Nation-State*, London: Longman.

Breuilly, J. (1993a) *Nationalism and the State*, Manchester: Manchester University Press.

Breuilly, J. (1993b) 'Sovereignty and Boundaries: Modern State Formation and National Identity in Germany', in M. Fulbrook (ed.) *National Histories and European History*, London: UCL Press, pp. 94–140.

Breuilly, J. (1996) *The Formation of the First German Nation-State, 1800-1871*, London: Macmillan.

Breuilly, J. (1997) 'The National Idea in Modern German History', in M. Fulbrook (ed.) *German History since 1800*, London: Edward Arnold.

Brubacker, R. (1992) *Citizenship and Nationhood in France and Germany*, Cambridge, Mass.: Harvard University Press.

Bulmer, M. and Rees, A. (eds) (1996) *Citizenship Today: The Contemporary Relevance of T. H. Marshall*, London: UCL Press.

Degler, C. N. (1992) 'One Among Many: The United States and National Unification', in G. S. Boritt (ed.) *Lincoln, the War President: The Gettysburg Lectures*, New York: Oxford University Press, pp. 93–119.

Demandt, A. (1990) *Deutschlands Grenzen in der Geschichte*, Munich: Beck.

Fahrmeir, A. (1997) 'Nineteenth Century German Citizenship: A Reconsideration', *The Historical Journal*, **40**(3), pp. 721–50.

Greengrass, M. (1991) *Conquest and Coalescence: the Shaping of the State in Early Modern Europe*, London: Edward Arnold.

Hobsbawm, E. (1991) *Nation and Nationalism since 1789: Programme, Myth and Reality*, Cambridge: Cambridge University Press.

Hobsbawm, E. (1994) *Age of Extremes: The Short Twentieth Century, 1914-1991*, London: Michael Joseph.

Hucko, E. M. (1987) *The Democratic Tradition: Four German Constitutions*, Leamington Spa: Berg.

Hughes, M. (1988) *Nationalism and Society in Germany 1800-1945*, London: Edward Arnold.

Jeismann, M. (1992) *Das Vaterland der Feinde: Studien zum nationalen Feindbegriff und*

Selbstverständnis in Deutschland und Frankreich 1792–1918, Stuttgart: Klett-Cotta.

Mann, M. (1996) 'Ruling class strategies and citizenship', in M. Bulmer and A. Rees (eds) *Citizenship Today: The Contemporary Relevance of T. H. Marshall*, London: UCL Press, pp. 125–44.

Marshall, T. H. (1950) *Citizenship and Social Class and Other Essays*, Cambridge: Cambridge University Press.

Rumpler, R. (1990) *Deutscher Bund und Deutsche Frage 1815–1866*, Munich: Oldenbourg.

Schulze, H. (1991) *The Course of German Nationalism: From Frederick the Great to Bismarck 1763–1867*, Cambridge: Cambridge University Press.

Sheehan, J. (1989) *German History 1770–1866*, Oxford: Clarendon Press.

Wehler, H.-U. (1987) *Deutsche Gesellschaftsgeschichte 1815–1845/49*, Munich: Beck.

Chapter 5

Austria: Nationality and the Borders of Identity

ANDREW BARKER

It was in the year 996 that the first recorded reference was made to a territory in East-Central Europe named Ostarrichi. 1996 thus marked the 1000th anniversary of the country we today call Austria, and such milestones provide a natural excuse not merely for celebration but also for stocktaking and analysis. While it may not figure at the centre of current debate, the Austrian example does, I believe, afford some instructive perspectives on issues of European integration and the relationship between national and supranational identities.

The shape of Austria today, a country whose borders have been more fluid than most, goes back to the post-First World War settlement subsequent to the collapse of the Habsburg Empire under the strain of modern political nationalism. At the same time, but in less dramatic circumstances, the United Kingdom of Great Britain and Ireland was truncated to form the United Kingdom of Great Britain and Northern Ireland. Both Austria and the United Kingdom thus assumed their present shape at more or less the same time and, like people in Britain, when Austrians contemplate possible developments within the European Union they can call on a long familiarity with multiculturalism and multinationalism. Indeed, since the collapse of the Habsburg monarchy, which saw itself as heir to the Roman Empire, the earliest of all European Unions, there has been an ongoing debate in both first and second Austrian Republics as regards political and cultural identity. Austria may be relatively new to the European Union, but its experience of debates about identity is old. It was more than mere coincidence that it was in the Austrian capital Vienna that Sigmund Freud conducted his experiments into the nature of individual identity.

Until 1918, the multi-ethnic Habsburg empire held a well-defined place on the geopolitical map of Europe. After centuries of mostly peaceful territorial expansion, its most recent and ultimately fateful acquisition had been Bosnia in 1878. It was of course in Sarajevo that the shots were fired which brought down not just the monarchy, but the whole European order. Gavrilo Princip's bullets effectively unleashed Europe's second 'Thirty Years War'[1] which the foundation of the European Community was primarily an attempt to settle. And if modern Austrians have (or have had) an identity crisis, this merely perpetuates problems going back to imperial times. With its German-speaking hegemony, the Dual Monarchy was an empire without an official name, but after the *Ausgleich*, the 1866 settlement with the Magyars, the shorthand names Austria-Hungary or simply Austria were most commonly employed. Voicing the truth of poetry, the novelist Robert Musil attempted to define what the empire had been:

> On paper it called itself the Austro-Hungarian Monarchy; in speaking, how-
> ever, one referred to it as 'Austria' – that is to say, it was known by a name
> which it had, as a state, solemnly renounced by oath while preserving it in all
> matters of sentiment, as a sign that feelings are just as important as constitu-
> tional law, and that regulations are not the really serious thing in life.[2]

Aside from the crucial observation that sentiment is just as significant as the legal process, certain historical parallels may suggest themselves here with the emergence of a multicultural new Europe, with the perceived threat to national sovereignty, the apparent disdain for national feeling and the quandary of forging a 'supranational' European identity out of constituent elements with strong national awareness. The Habsburg Empire had already tried to create just such a supranational unit, but it foundered not just on the demands of the Czechs, Slovaks, Magyars and others for political autonomy, but also on the demands of German nationalism. For much of this century, Austrians living in the rump republic, the vast majority of them German speakers, have had to cope with the question of their identity first in the aftermath of the loss of an empire which they had dominated culturally and politically, then in the context of National Socialism which forcibly placed them into the German camp, then in the post-war settlement which just as determinedly tried to take them out of it, and finally now in the 'New Europe'.

Writing in 1917, as the war-ravaged empire began to disintegrate, the Viennese poet Hugo von Hofmannsthal developed the notion of an 'Austrian idea' which, through the coexistence of many differing peoples within one political structure, might be held up as a microcosm for a Europe of the future. Hofmannsthal was firmly of the view that the shared cultural heritage of the various constituent peoples of the Danube basin was infinitely more significant than any political structure. In this he anticipated Jean Monnet, who eventually concluded that emphasizing shared cultural values and achievements might have provided the best of all launching pads in the movement towards the ultimate goal of a true European union. Falsely assuming that the Austrian Empire would remain intact after the war was over, Hofmannsthal claimed that the new Europe which would emerge required the specific example of Austria and what he called its 'unforced elasticity'. Thus Hofmannsthal and other like-minded Austrian intellectuals and artists helped found the Salzburg Festival directly after the First World War in what was very much an attempt to demonstrate how the specific and the local, that is, the city of Salzburg itself, could also be the focal point for the projection of an ideal which stressed the oneness of European culture.

In Central Europe the Habsburgs' long-term failure to provide a pluralist solution within a single political unit was forgotten surprisingly quickly. Indeed, within a few years their failed attempt at supranationalism was being nostalgically recalled, above all by German-language Jewish writers of the 1930s like Joseph Roth, who hailed from Galicia, the Moravian Ernst Weiss and the Viennese Stefan Zweig. The Habsburg past, viewed in the light of the Hitlerian present, was now recast as what Zweig recalled as a 'Golden Age of Security.'[3] That the National Socialism which threatened them was itself a product of the recent Habsburg past, which these writers now glorified, was an irony they would have preferred not to acknowledge. All three committed suicide in the late 1930s and early 1940s.

After 1918, when the original ambition of the new Austrian Republic to style itself *Deutsch-Österreich* had been thwarted by the international community, there developed a twin approach to the problem of role/identity in the new Austrian Republic. The first stressed the common ethnicity of Germans and Austrians and developed the dream of union with Germany. This was the Anschluss so fervently desired by both nationalist Right and internationalist Left. When the enforced union came in March 1938 it was genuinely popular, even though the 99 per cent 'Yes' vote in the ensuing plebiscite has to be taken with a pinch of salt.[4] The second approach was to develop the notion of an 'Austrian' identity distinct from a German one. This movement was very much led by writers and intellectuals such as Hugo von Hofmannsthal, Josef Redlich and Leopold von Andrian, calling on a tradition which itself went back at least to the earlier part of the nineteenth century. Then too, it had been creative writers who had articulated the need to distinguish between the German and the Austrian. In 1837, for example, the outstanding Viennese dramatist Franz Grillparzer wrote an essay entitled 'Worin unterscheiden sich die österreichischen Dichter von den burden' ('How are Austrian poets different from the others') in which he concerns himself exclusively with delimiting the specifically Austrian from the German scene as a whole. It was this latter approach which, understandably enough, received the fervent backing of the reconstituted Austrian republic in the years after 1945.

The newly restored Austrian state faced a problem determined at least as much by language as it was by history. The Anschluss mentality of so many Austrians was encouraged by the resonance of the very word *deutsch*, which is both the name for a language and a conditioner of identity in a way which the word English, to take the obvious example, is not. The point is simply made: there is a world of difference between being a native speaker of English and being English. Being anglophone is not a determinant of national or even cultural identity. With German this division between language and identity is far less clear. Even the proudest and most independently minded Bavarian is also *deutsch*. This condition was, and to a diminishing degree still is, equally present in Austria, where even the pan-European Hofmannsthal was unable to break out of a linguistic bind which goes back to the very meaning of *deutsch*, a word cognate with the words signifying clear or understandable. To speak *deutsch* means simply to speak in an intelligible fashion. We even have a remnant of this usage in the English language, where to speak Double Dutch is to talk nonsense. The notion of *Deutschland* therefore was originally no more than shorthand for the geographical area in which people lived who could communicate linguistically with one another. Things looked pretty much the same in Eastern Europe, where the general Slavic root for the German they could not understand, *nemetz*, means deaf and dumb. The notion of Deutschland as a political entity came much later, conditioned to a large degree by nineteenth-century thinking on biological nationalism. Significantly, Hoffmann von Fallersleben's poem and later anthem *Deutschland, Deutschland über alles* was written at a time when there was no such thing as the German state. Hence even Hugo von Hofmannsthal (1874–1929) believed that in the last resort it was impossible to be Austrian and not also German. His friend Leopold von Andrian (1875–1951), on the other hand, believed it was possible. Unable to overcome the dual identity problem, the 'solution' adopted by Hofmannsthal after the mid 1920s was to move away from the 'Austrian idea' to the notion of an overarching 'European Idea', understood primarily in cultural rather than political terms.

The period 1938–45, when an independent Austria was absorbed into the Third Reich, saw the logical conclusion of the notion of being 'Austrian' as a sub-category of 'German' identity. Hitler himself was famously an Austrian who thought of himself as a 'German'. The Jewish Viennese writer Alfred Polgar concluded that the difference between Austrians and Germans was that the 'Germans are first-class Nazis, but lousy anti-semites; the Austrians are lousy Nazis, but by God what first-class anti-semites they are!'[5] Nevertheless, the residual power of an 'Austrian' identity substantially discrete from a German one was illustrated above all by the Nazis' immediate proscription of the very word 'Austria' on seizing power in March 1938. The ancient provinces of Upper and Lower Austria were at once renamed Upper and Lower Danube, while Austria itself became the *Ostmark*, or Eastern Marches.

After the war the decision was taken at the highest level to restore the Austrian State to its 1938 boundaries, a process initiated by the benevolent Moscow Declaration of 1943 which maintained the fiction that Austria, far from being an often enthusiastic collaborator in an ideology born in the cafés and debating societies of Vienna, had in fact been the 'first victim of fascism'. From this conscious wartime decision to re-establish or recreate an Austrian state distinct from a 'German' one there has gradually developed a situation in which there are now many individuals who can and do separate their being speakers of German from their identification with a German nation requiring German statehood to validate it. It is a striking example of what Eric Hobsbawm famously called 'the invention of tradition'. To a quite large measure this creation of identity has been more successful than the development of a British identity from the eighteenth century onwards. Here it has always remained the case that the supranational British identity went hand in hand with allegiance to Scotland, Ireland, Wales or England. In Austria most people today who still consider themselves as part of what can be termed the *deutsche Kulturnation* would place their Austrian identity before their German one. In Scotland today, I suspect, Scottish identity would for most people come before Britishness.

However, the case for an Austrian national identity remains a problem for some, and not just older, citizens of the Second Republic. This is most evident in the startling success of a young[ish] politician like Jörg Haider, leader of the Austrian Freedom Party (FPÖ) which was traditionally the last resting ground for old Nazis and is now the spiritual home for new ones too. Especially problematic has been the fact that since 1945 the success in creating an independent Austrian identity lay to such a large extent in the conscious suppression of Austria's own darker history. Unlike the Germans, Austrians were encouraged to believe that there was no problematic past with which they had to come to terms (the German expression is *Vergangenheitsbewältigung*). This repression of the past rebounded viciously with the Kurt Waldheim affair of the mid-1980s. Then it emerged just how much the development of an independent identity had been at the cost of an honest assessment of the Austrian complicity in Nazi atrocities.

Once again writers and intellectuals had been crucial players in the gameplan to create an 'Austrian' identity: in his novels of the early 1950s, for example, the reformed Nazi Heimito von Doderer (1896–1966) had played a crucial role in persuading both intellectuals and the wider literate public of the importance of their specifically Austrian heritage. The essence of his work, however, was that the years 1938–45 had somehow been an aberration, and that the new Austria should seek to restore the historical continuum of the Habsburg Empire and the First Republic, destroyed by the intervention from outside of National Socialism. Like many other latter-day

Austrian patriots, Doderer only began to appreciate his Austrian identity once it was proscribed by the very organization which he had joined when it was illegal in Austria to do so and whose aim was to subvert the struggling First Republic.

The extent to which the furtherance of the new Austrian state had been based on a mixture of political amnesia and selective history only emerged in the wake of the Waldheim scandal. Once again literature was at the fore in the debate about national identity. Now it was the turn of the novelist and dramatist Thomas Bernhard (1931–89) to excoriate the duplicity of the attempt to create an Austrian identity by pretending that a certain segment of the past had not really happened. In the novel *Auslöschung* ['Obliteration'] and the drama *Heldenplatz* ['Heroes' Square' – the vast space in front of the Imperial *Hofburg* where Hitler addressed a vast and adoring multitude soon after the take-over of 1938], Bernhard directed a stream of inspired bile against his own country possibly unmatched in the literature of any other European nation.

Far from it being another and distant country, the past assumed enormous proportions in the debates about Austrian identity and culpability which were unleashed by the Waldheim affair. It is perhaps no coincidence that successive Austrian governments became such enthusiastic supporters of Austria's accession to the European Union. For in the vision of a new, possibly federal, Europe there presents itself a golden opportunity for the country to be at one and the same time aware of itself in its Austrianness, in the context of its wider German ethnicity and in the sense of its historical experience of life in a multi-ethnic community. The eventual expansion of the Community to include countries which earlier this century were still part of the Habsburg realms is welcomed by many Austrians who see the country linking up again with its own past. Moreover, it is interesting to note the resurgence of nostalgia for the past in the successor states to the empire. A good example would be the continuing popularity in southern Poland of the 'Habsburg idea' of a Catholic, multicultural association of states whose golden, or rather muddy brown, thread is the Danube.

However, before being transported by visions of a future in which former secessionists regroup, forming a new order apparently replicating aspects of a corporate past once strenuously rejected, it is important to remember that although Austria, like Germany, now has fixed and accepted geopolitical boundaries, some internal psychological boundaries remain. And they are problematic. It may be that more will yet be heard of Jörg Haider, or Haider Jörg as he is known in dialect, a rabble-rouser who is more than happy to quietly exploit the parallel between his initials and those of an earlier HJ – the Hitler Jugend.

Notes

1. Mayer (1981), p. 3.
2. Janik and Toulmin (1973), p. 36.
3. Zweig (1943), p. 13.
4. Rathkolb *et al.* (1990), pp. 15–16.
5. Clare (1981), p. 188.

Bibliography

Barker, A. (1996) 'Doderer's Habsburg Myth: The Novel, History and National Identity', in A. Bushell (ed.) *Austria 1945–1955. Studies in Political and Cultural Re-Emergence*, Cardiff: University of Wales Press, pp. 37–54.

Clare, G. (1981) *Last Waltz in Vienna. The Destruction of a Family 1842–1952*, London: Macmillan.

Frank, P. and Pörnbacher, K. (1964) *Franz Grillparzer. Sämtliche Werke*, Volume 3, Munich: Carl Hanser, pp. 809–11.

Hobsbawm, E. and Ranger, T. (1983) *The Invention of Tradition*, Cambridge: Cambridge University Press.

Janik, A. J. and Toulmin, S. (1973) *Wittgenstein's Vienna*. New York: Simon & Schuster.

Mayer, A. J. (1981) *The Persistence of the Old Regime. Europe to the Great War*, London: Croom Helm.

Rathkolb, O., Schmid, G. and Heiß, G. (1990) *Österreich und Deutschlands Größe: Ein schlampiges Verhältnis*, Salzburg: Otto Müller.

Zweig, S. (1943) *The World of Yesterday*, London: Cassell.

Chapter 6

Perspectives on Frontiers: The Case of Alpe Adria

———

RAIMONDO STRASSOLDO

A geo-historical introduction: frontier problems in North-Eastern Italy

When in 1968, in the Italian frontier town of Gorizia, a research institute was established with the specific aim to study the problems related to borders, there was no body of Italian social-scientific tradition on this matter to rely on, and very little material internationally (with the exception of political geography).

Borders and frontiers still do not appear to be a relevant topic for social-scientific inquiry in Italy. One reason may be that most of Italy is a peninsula, and the coasts are not commonly conceived as borders; most of the land borders run on the crest line of the High Alps and are therefore thinly populated. Physical contact between Italy and the rest of Europe takes place in a relatively limited number of border passes, where specialized settlements have developed. Italian alpine borders are characterized by remoteness, marginality, peripherality, out-migration (except where tourist development obtains) and by a series of often highly congested, mostly minor border towns. It seems that this dual nature has hindered the development of notions of borderlands and the 'border situation' as a unified problem worthy of scientific study. Also, the notion of 'frontier region' is problematic, since most of Northern Italy would fall into this category: all administrative regions north of the Po river extend to the state boundaries, but this hardly characterizes their identity.

The case of South Tyrol

The three Italian regions where the border is a relevant, even a central, problem, are Valle d'Aosta, Trentino-Alto Adige (South Tyrol) and Friuli-Venezia Giulia. In the first two cases, although the border runs along the highest alpine peaks, the state boundary cuts across regions unified from time immemorial, and it separates two halves of a unitary language community. In the case of South Tyrol, the frontier was imposed by military force in

1918, well beyond the expectations and claims of the Italian 'irredentist' groups; it has been highly resented, and never completely accepted, by the local population. After 1945, and after the bombings and killings of the fifties and the sixties, Italy could keep this boundary only at the price of granting South Tyrol a high degree of autonomy and many financial concessions. Nevertheless, problems of inter-ethnic relations, of federalism, of transfrontier co-operation, of border infrastructures and other typical issues of the frontier *problèmatique* continue to be at the centre of both political and social-scientific interest in this region. For example, recent (1995) plans for closer co-operation between North and South Tyrol were heavily handedly vetoed by Italian national authorities as getting too close to secession. Incidents like these stimulate meetings, conferences, debates, studies and publications which can be classified as belonging to 'frontier literature'.[1] Some of the same interests and outlooks have permeated the neighbouring Italian province of Trento, which has been very much influenced by the vicissitudes of South Tyrol; it has shown strong autonomist aspirations, stresses its 'Central-European' ties and ethno-regional peculiarities.[2]

The case of Friuli and Venezia Giulia

In Friuli-Venezia Giulia, the situation is even more complicated and, in the past, has been more tragic. This is the only place in Europe where the three main European culture areas – German, Latin, and Slav – meet, and have done so since the seventh century. Central European powers (in particular the Habsburg Empire) have here confronted the Mediterranean powers: especially Venice, and then Italy. In the easily passable Eastern section of the Alps, from Tarvis to Gorizia, the frontier between these powers has frequently been disputed, and subject to drastic relocation as a consequence of wars. For many centuries (1420–1866), it cut across the same populations, neo-Latin Friulians in the plains, and Slovenes in the highlands. In recent times, this has produced a problem of national minorities (Slovenes and some Germans on the Italian side, Italians on the Austrian and then Yugoslav side). As a consequence of victory in 1918, the Italian state pushed the boundary deep into ethnic Slovene territory. When the fortunes of war were reversed, in the late stages of the Second World War, Tito's Yugoslavia claimed half of Friuli up to the Tagliamento river, on the ground of ancient Slovene settlements, plus the two cities of Gorizia and Trieste. This led to bloody conflict within the anti-Nazi partisan forces, between the Communist/Slovene and their Italian Communist supporters on one side, and the rest of the Italian resistance on the other. Fascist policy of brutal repression and forced assimilation of the Slovene minority had built a deep hatred of Italians, which in 1945 flared into mass murders, genocide and 'ethnic cleansing' in the Tito-occupied 'Venezia Giulia' (Gorizia, Trieste, Istria and Dalmatia). Several thousand Italian civilians were horribly killed in Karst caves (the 'foibe') and about 350,000 fled their Yugoslav-occupied homelands. Conflict continued in the following years over the status of Trieste, and until 1953 Italy and Yugoslavia were rattling sabres. The problem of the north-eastern boundary was for many years one of the main focuses of Italian politics, both internal (it became the test of national dignity for the new democratic republic) and international. In 1954, a 'temporary' agreement called the London Memorandum was reached, but only in 1975 a final peace treaty between Italy and Yugoslavia was signed – the Treaty of Osimo.

The normalization of cross-border relations in the Upper Adriatic

For ten years – from 1945 to 1955 – the de facto boundary between Italy and Yugoslavia, running a few kilometres east of Trieste and through the town of Gorizia, was effectively sealed, and formed part of the Iron Curtain; the long conflict – from 1918 to 1954 – over minorities and territory, with its massacres, had left a legacy of deep suspicion and hate. After 1955, neighbouring relations were very cautiously resumed, mainly under the pressure of local economic needs (cross-border property rights, primary supplies, etc.). In the 1960s, with the growth of private motor car ownership and the receding of war memories, that border traffic began to grow, with Italians crossing into Yugoslavia to take advantage of the much lower prices there, particularly petrol and meat, and the Yugoslavs, in turn, buying manufactured goods (mainly clothing and home appliances) in Italy. Slowly, tourism drew adventurous Italians into the Alpine and coastal resorts of Slovenia, Istria and Dalmatia.

By the late 1960s, a new generation had matured, which had not personally experienced the horrors of Fascism and war. New attitudes towards the neighbours on the other side of the border developed. These new attitudes were shared by the political class; transfrontier contacts between local authorities started again. Common interests in the economic sphere were discussed, and also common social ties and cultural values. Cross-border relations became 'civilized' again.[3]

The role of the region Friuli-Venezia Giulia

An important stage was the institution, in 1963, of the autonomous region Friuli-Venezia Giulia which set itself the task of becoming the 'bridge' between Italy and its eastern neighbours, beginning with the Yugoslav federal republics of Slovenia and Croatia. The Austrian *Länder* of Carinthia and Styria were also identified as partners in cross-border co-operation. Thus the region Friuli-Venezia Giulia started to develop an inter-regional, international policy of its own – informally, since its statutes and the Italian Constitution did not allow such activities. One of the means by which such policies were pursued was the establishment of semi-private institutions. Another was involvement in European initiatives – the Council of Europe and the EEC – in the field of cross-border co-operation, and in what has been called the 'European Frontier Region Movement'.

The Institute of International Sociology of Gorizia was one of those institutions; but there were others, like the Institute for Central-European Cultural Meetings, established in Gorizia in 1966, which revived contacts between intellectuals and artists of the area of the former Habsburg Empire which, at the time, mostly belonged to the Soviet 'empire'; the Regional Institute for European Studies which acted more on the middle-brow and popular-culture level, promoted European consciousness, values and knowledge; the Institute for the Study and Documentation on East Europe specialized in gathering, processing and distributing information on the economic developments in the area; and others. The region also patronized more contingent and special initiatives of a cross-border and inter-regional nature, thus strengthening the international outlook of the regional community: meetings of local authorities, conferences of special professional groups and interests, sports and cultural events, twinning of municipalities, etc.

Border studies at the Gorizia Institute of International Sociology (ISIG)

These activities formed one of the fields of research of ISIG; the second main interest was the study of inter-ethnic relations in this and other border areas. The early publications of the Institute include a theoretical-programmatic statement,[4] a statistical-economic analysis of border traffic in Gorizia,[5] a study of the technical-legal aspects of Italian boundary controls,[6] and a historical-geographical study of the complex vicissitudes of Italy's northeastern boundary.[7] The psychological, cultural and social aspects of 'living at the border' were the topic for a properly sociological field research – a sample survey on 1,215 respondents from the Gorizia and Trieste area – carried out by Renzo Gubert in 1972.[8] A study was done on a feature typical of conflictual frontier areas, that is, the militarization of the territory.[9] This first wave of activities culminated in the calling of an international conference of experts in various social science disciplines concerned with border problems.[10]

Meanwhile, the expertise developed at ISIG in border-related problems was called on by both regional and European bodies. Researchers from the

Institute assisted the region in drawing up documents on cross-border co-operation, and acted as consultants to the Council of Europe in developing activities on behalf of frontier regions.[11] The study of border problems was then pursued at a more theoretical level.[12] The study of cross-border activities in the area of Friuli-Venezia Giulia continued into the 1980s, particularly by the work of Giovanni Delli Zotti.[13]

In the following years, research projects on ethnic minorities – a common feature of border areas – attracted the most attention. A sample survey of attitudes, perceptions and stereotypes among eleven ethnic communities (or sub-communities of Latin and Slavic stock) living along the Italian side of the Italian-Yugoslav border was carried out in 1973, although it was only published eight years later.[14] In the same year, another sample survey was conducted by ISIG in a multi-ethnic area of Trentino-South Tyrol.[15] A textbook on ethnic relations, a consequence of these research interests, was published, which for a long time remained the only book in this field available in Italian.[16] ISIG also assisted Professor Feliks Gross of New York City University in a study on border-ethnic problems in this region.[17] This emphasis on ethnic issues characterized the second main ISIG conference on border problems, organized to mark the tenth anniversary of the founding of the Institute. The proceedings were published in two volumes (in English), one on various aspects of co-operation and conflict in border areas,[18] and the other on ethnic minorities in the borderlands.[19] In the 1980s and 1990s, several studies on ethnic groups, minorities and language groups were undertaken at ISIG.[20]

The development of cross-border, inter-regional co-operation in the Alpe Adria Area

One of the main objects of study, and one of the main sponsors of studies on border problems was, in the 1980s, that entity called Alpe Adria. Alpe Adria is one of the 'working communities' formed by regional and local authorities along European frontiers. It first appeared as 'Trigon', a private, informal group of regional planners of Friuli-Venezia Giulia, Carinthia and Slovenia, meeting in the late 1960s to arrive at common ideas on the infrastructural and economic development of the area. The improvement of road and rail connections between the Danube basin and the upper Adriatic, overcoming the Alpine barrier, was the basic issue. Soon Croatians joined the group (now re-christened 'Quadrigon'). At the same time, a variety of private and semi-public bodies (like universities and chambers of commerce) and local authorities promoted their own cross-border links.[21]

The need for more orderly institutional arrangements was felt, and the Regio, Euregio and Arge-Alp examples were at hand. About ten years after the first beginnings, the Working Community Alpe Adria was officially christened in Venice in 1978. Its very name, echoing the word *Arbeitsge-meinschaft*, stresses the important role played by the German partners as midwives. Although Bavaria participated only as an observer, albeit an active one, it was to be one of the most significant and most involved partners of Alpe Adria, mainly because of its need to improve connections with the Adriatic harbours (Venice and Trieste, but also Koper/Capodistria and Rjeka/Fiume). The original full members, besides the already mentioned Friuli-Venezia Giulia, Carinthia, Slovenia and Croatia, were the Austrian *Länder* of Salzburg, Upper Austria and Styria, and the Italian region of Veneto. The working programme included the setting up of a series of working committees for specific problems; the first was concerned with regional planning and environmental management. The others dealt with transport, culture, science and sports, economy and tourism, agriculture, forestry, animal production and mountain economy, health, and social affairs, respectively. Each committee was charged with establishing specific objectives, methods and schedules, and presenting results in the form of reports. These have usually an analytical-descriptive part, presenting the state of the question in each member region, and a policy-orientated part, commenting on the differences between the regions, recommending strat-egies for the harmonization of policies and setting common goals. Some of these reports were given wide circulation in the form of handsomely illus-trated documents and books. Other public activities of Alpe Adria took the form of promotional events and exhibitions. Periodically, the senior political authorities of all member regions would meet in plenary sessions to discuss and approve the work done, work out new projects and issue high-sounding public declarations.

The organizational infrastructure supporting this work was, and remains, rather scanty. There is no permanent secretariat; Alpe Adria functions as a network of officials in each regional government. Until 1991 (Declaration of Linz), there was no common budget. Each member region would bear the costs of their own activities for and on behalf of Alpe Adria. An elaborate rotation system was adopted to share responsibilities and tasks. Each region was asked to play the leading role in each project for a certain time: it would act both as chair and as 'local organizing committee' for meetings, agendas, hosting, etc. Meetings took on all the formal features of diplomatic events, with strict observance of rules regarding the use of languages, precedents, etc.

In a short time, outer layers of regions applied for admittance to the original group. To the west, Alpe Adria incorporated Trentino-South Tyrol, Lombardy and the Swiss canton of Tessin; to the east, Austrian Burgenland, and then the Hungarian counties of Györ-Sopron, Vas, Zala, Somogy and Baranya. Talks were also begun to negotiate admission for some areas of

Czechoslovakia. Thus, a sizeable part of Central Europe seemed to be organizing around Alpe Adria.

It is hard to tell what would have become of Alpe Adria if it had been permitted to develop along the lines set in the first ten years of its life. To expand from nine to nineteen regions, from four to eight state systems, and from four to seven different languages, makes co-operation a difficult task. This is especially so considering that most of the work had to be done outside the formal legal competence of the regions involved. Most of the regional governments active in the Alpe Adria set-up had no statutory powers to do so; only Bavaria, Slovenia and Croatia had, to a limited extent. Austrian *Länder* had to wait until 1989 for constitutional amendments which would empower them to do what they had been doing for many years. Italian regions, to this day, have no powers whatsoever in the international field.

A central question concerns the practical effects of this activity. Somewhat cynically, it could be maintained that it amounts to mountains of printed paper – technical reports, statistical analyses, glossy promotional picture books – endless streams of political rhetoric and a plethora of meetings of politicians and officials in luxury hotels located in attractive tourist resorts. Indeed, the translation of all this activity into concrete legal changes and administrative decisions in each region seems to have been small, if at all.

But Alpe Adria managed to become a reality in the consciousness of ordinary citizens. Many enterprises, straddling borders in this area, have borrowed the name – for example, the motorway linking Friuli to Austria, radio stations, shopping centres, cultural associations, residential developments. A project is in train to have the International Olympic Committee design the *Dreiländereck* of Tarvis, Villach and Kranjska Gora as the venue for future Winter Olympics the first Olympic Games jointly hosted by three countries. The promoting committee failed the 2002 target but is trying again for 2006.

Perhaps more important, Alpe Adria has produced a feeling of mutual knowledge and understanding, of goodwill and community among the highest officials and political leaders of the area.[22] This has undoubtedly helped to ease the solution of concrete problems occurring between them such as, for instance, when Austria enforced a restrictive policy on commercial transit-traffic on its routes.

The role of Alpe Adria after the 1989–1991 revolution in Central Europe

Perhaps the most dramatic example of the concrete effects of the Alpe Adria co-operation was the unhesitant solidarity that the Italian neighbouring regions, and especially Friuli-Venezia Giulia, offered to Slovenia and Croatia during the critical weeks of the breakaway from Yugoslavia in 1991. In contrast to the cautious and conservative pro-Belgrade, pro-Yugoslav policy of the Italian central government, the regional authorities of Friuli-Venzia Giulia quickly sided with Slovenia's and Croatia's bid for independence. It was widely acknowledged that this 'scandalous' difference between the central and the regional position on an international issue was largely due to the long experience of co-operation within the Alpe Adria community.[23]

After 1989 and 1991, Alpe Adria underwent a period of uncertainty, which it has not yet overcome. The future is unclear, because the general political situation has fundamentally changed. One of the aims of Alpe Adria was to devise ways of practical co-operation among regional communities belonging to three different socio-economic-political systems – Western capitalism and liberal democracy, Yugoslav one-party self-management and Hungarian 'gulasch-socialism'. Since 1989–91, the latter two have disappeared; the former has become the system common to all regions of the area. In principle, co-operation could now be based on more traditional, formal, state-led channels. This has led to the launching of the so-called Central European Initiative, of which more below.

The second crucial change is that, after 1991, two of the member regions, Slovenia and Croatia, graduated into fully sovereign nation-states. This makes it awkward for them to keep their membership in an organization of sub-national entities. After independence day, Slovenia and Croatia vowed to keep their membership of Alpe Adria, in gratitude for the solidarity received from other members; but they would participate at the level of Foreign Ministers, not of Heads of State.

The third development is the integration of Austria into the European Union since January 1996 which has changed the character of the Italian-Austrian border from an 'external' to an 'internal' EU frontier and, in turn, has transformed Austria's borders with its neighbours to the west and east into external frontiers of the EU. This may be a temporary situation, since the extension of EU membership to the Czech Republic, Slovenia, Hungary and, probably a little later, to Slovakia and Croatia are on the European agenda. Yet all these changes in status of the borders are bound to have many practical consequences on border relations in the area.

The Central European Initiative and the revival of nationalisms in the Alpe Adria region

In the later 1980s the idea of Alpe Adria – whatever its real substance – seemed to be spreading into Central Europe, coalescing members from Lake Maggiore to the Balaton. With the dissolution of the Soviet empire, the opportunity arose for central governments of the area to step in and resume the leading role in these activities. Largely under the prodding of the Italian government, in particular by Foreign Minister de Michelis, the idea of some sort of intergovernmental community-building in this area took form. The result was something called, first, the 'Quadrangle' (1989), comprising Italy, Yugoslavia, Austria, and Hungary, then 'Pentagon', adding Czechoslovakia, then 'Hexagon', when Poland joined, before it became finally known as the Central European Initiative. Other countries, like Belarus, Romania, and Bulgaria, expressed interest in an association. One of the first acts under this initiative was the Millstat Declaration (1991) in which the member states voiced, among other things, their appreciation and support for co-operative activities at the inter-regional level – such as Alpe Adria. However, it was clear that central governments intended to take the lead in this field. The need for autonomous, spontaneous initiatives of the regions was now less pressing, and initiatives such as Alpe Adria were jeopardized.

The second development was the revival of nationalist and right-wing attitudes in most countries of the Alpe Adria area. The roots of this phenomenon need not be discussed here, and are different in each country. Suffice it to note that in the 1990s they have seriously affected bilateral relations between Italy, Slovenia and Croatia. In 1994, the new centre-right coalition in Rome revived controversy with the Yugoslav successor states, requesting a revision of the Treaty of Osimo, especially on the points concerning the property rights of Italian refugees. The equally strongly nationalist governments in Slovenia and Croatia resisted, and Italy brought the dispute to the European level, vetoing Slovenia's association agreement with the EU. The old questions of the status of the Slovenian minority in Italy and the Italian minority in Slovenia and Croatia were also revived. Thus, intergovernmental relations between Rome, Ljubljana and Zagreb reverted to levels of tension almost as high as in the 1950s. The regional government of Friuli-Venezia Giulia made it clear that it did not agree with Rome's hard-line approach but, unavoidably, the inter-state tensions rebounded on transfrontier relations and on the working of Alpe Adria.

The re-emergence of the Istria question is connected with the dissolution of Yugoslavia. What had been an internal, administrative, invisible line between the federal republics of Slovenia and Croatia became a fully fledged, tightly guarded international boundary between two sovereign states. Among other consequences, the new boundary cut the Italian minority into two halves, with different legal status. The minority in the part now belonging to Croatia had many reasons for concern, in the face of the nationalist, centralist and authoritarian Tudjman regime. For this and other reasons, the idea spread among Istrian intellectuals of claiming for Istria a special status, with international implications. Taking the lead from the Tyrolean idea of integrating Austria's north Tyrol and Italy's South Tyrol within a single 'Euregio Tyrol', the suggestion was made for a similar status for Istria: 'Euregio Istria', with complex and somewhat nebulous ties to all three states concerned – Slovenia, Croatia and Italy. This has stirred up heated discussion,[24] and causes deep suspicion in Ljubljana and Zagreb, ever fearful of Italian revanchism. However far-fetched Italian revisionist claims may be, the Istrian population manifests growing opposition to Tudjman's regime; Istria is trying to revive what little is left of its Italian heritage, and to resume relations across the Adriatic with Venice.

The consequences of 1989 on Italian internal politics: the emergence of new autonomist movements in Italy's northern regions

The sudden collapse of Communism in eastern Europe had seismic consequences in Italian internal politics. The Italian Communist Party (with about 30 per cent of the vote the largest Communist party in western Europe) finally repudiated Communist ideology, changed its name to Democratic Party of the Left, and ceased to appear as a threat to the liberal-democratic-capitalist system. In turn, the parties which based their strategy on the opposition to Communism lost one of their main functions. In conjunction with many other factors, this led to the emergence, in Italy's most developed northern regions, of new political formations whose main goal was the acquisition of much greater regional autonomy, and the transformation of Italy from a centralist-unitary state to a federal republic. Such movements had already existed for some time at the margin of the established party system, in the regions of Friuli (*movimento Friuli*) and Veneto (*Liga Veneta*). At the beginning of the 1980s, the *Lega Lombarda* was born, and at the end of the decade it benefited enormously from the collapse of Communism. After 1991, it also benefited from the exposing of the widespread corruption of the old party system ('Operation Clean Hands'). Within a few years, all regions north of the Po river were affected by autonomist-federalist movements, eventually brought together into the 'Northern League'. Almost one-third of the moderate, centrist electorate abandoned the old parties and switched their allegiance to the League. The level of support for this movement was directly and strongly correlated to latitude – northern location – and proximity to the Alpine border.[25] The April 1996 elections showed that the phenomenon had established solid, stable roots, especially in the north-eastern regions of Veneto and Friuli.

The factors explaining the rise of the League are numerous and complex. Some of them undoubtedly originate in the external political environment. The League can be seen partly as a response to the stresses and opportunities of the European integration process: the developed Northern regions, already well integrated economically into Europe, fear that the backward South would hold Italy back and make it drift into the Mediterranean, African world – they see 'separate development' as their opportunity to avoid that fate. The League has also profited in many ways from the dissolution of the Communist bloc; not only, as already mentioned, from the disappearance of the internal 'Communist threat', but also from the emergence of 'new-old' nations from the old state shells. The example of Bosnia was, of course, a deterrent; but the Baltic countries, Slovenia, Croatia and, eventually, Slovakia showed that intangibility of boundaries and State self-preservation were no longer sacrosanct, and (sub-)national self-determination no longer just a dream. This progressively moved the League's ideology from regional autonomism to federalism to mini-nationalism (the 'Northern Nation') and, eventually, to demands for independence, separation and secession.

This shift is mostly tactics and rhetoric; but it seems that the drive to a greater degree of self-government in the northern – and especially in the north-eastern – regions is gaining momentum. This can be explained by their geography and cultural history. Autonomist aspirations are strongest along the borders because the people living there have a long history of contacts and exchanges across these borders. Lombards are familiar with the Swiss federal system and see its advantages. South Tyroleans, of course, identify much more with the North Tyrolese than with their fellow-Italians; people in Trentino, too, since the province had been part of the Habsburg Empire, seem to be culturally orientated more towards the North than to the rest of Italy. Veneto's case is different; its autonomist feelings seem to be nurtured more by economic factors – fiscal revolt, complaints about deficiencies in the State's infrastructure, a regional economy strongly export-oriented – and by ethnic prejudice against Southerners than by cultural-political reference to the old, glorious Venetian Republic. In addition, the 'border' character of Veneto is in fact negligible.

Federalism and autonomism in Friuli-Venezia Giulia

By contrast, Friuli-Venezia Giulia is decisively marked by its location on the border. From the beginning of time, the region has been moulded by that fact. Ethnically, it is the result of a complex web of relationships between the three main peoples which meet in this corner of Europe – the Latins, the Germans and the Slavs. Economically, it has lived, in the non-agricultural sectors, mostly from trade with Central Europe. Culturally and politically, its history has been patterned by the presence of a military frontier between the Italian (formerly Venetian) and the Central European powers, over which many wars, some of them major, have been fought.

In recent decades, history seems to have been diverted from its bloody course. Friuli-Venezia Giulia has begun to see itself not as a bulwark nor a battlefield, but as a busy bridge between Italy and its northern and eastern neighbours, as active part of a network of peaceful relations between the upper Adriatic and the Danube basin. To develop this role, Friuli-Venezia Giulia claims more freedom of action and a greater degree of self-government. The long experience of co-operation in the Alpe Adria context, with partners belonging to federal states, has exposed the regional political class to the advantages of such systems. Older, marginal autonomist movements, based mainly on ethnic-regional, inward-looking, local concerns, have merged into the largest political power in the region: about 25 per cent of the vote has recently gone to the Northern League. The regional government of Friuli-Venezia Giulia has been, since 1994, the only Italian region headed by the League. Following Mr Bossi's federalist strategies – but showing some caution about his recent demagogic utterances about independence and secession – the Friulian League has developed plans for more regional autonomy, including authority over international, inter-regional and cross-border relations. Other political groupings have done the same. Almost everyone – even the right-wing parties – demands more autonomy, a stronger regional identity and increasing integration into Europe; almost everyone in democratic politics points to the Swiss, German and Austrian federal experience as positive models which Friulians are able and well-qualified to follow.

It is difficult, even impossible, to predict future developments. The Italian political system is undergoing a deep transformation, and is still far from having reached a new equilibrium. Events in Friuli-Venezia Giulia continue to depend, above all, on what happens in the rest of the Italian state system. However, the drive for decentralization and federalism in the rich north-eastern regions seems unstoppable, and all Italian political forces agree that this political claim must be in some way satisfied. One question is whether Friuli-Venezia Giulia will be able to maintain a separate identity and autonomy, or whether it will merge with the larger 'Padania' or 'Triveneto' macro-regions – as envisaged in the plans of the Northern League. Its peculiarities as a border region will probably be impossible to suppress. The domination of Milan, Mantova or Venice will not be more acceptable than Rome's.

But this prognosis applies mainly to Friuli itself. Venezia Giulia and Trieste's tiny territory are different in history, character, interests and political orientation. Trieste is still characterized by century-old anti-Slav feelings, heightened by the events of 1944–7, and thus has always been a stronghold of right-wing nationalist parties. The population of Venezia Giulia is less keen on regional autonomy; its interest lies in privileged relationship with Rome (before 1918, with Vienna). Trieste is traditionally interested not so much in border relations and good neighbourliness, but in what the Germans call *Grossraumbeziehungen*, in spatially broad and long-range relations. Although the administrative capital of Friuli-Venezia Giulia, Trieste's nationalist and refugee lobby was the main force behind the recent difficulties in the relationship between Italy and Slovenia. The regional government, traditionally headed by Friulians, and its policies towards the eastern neighbours are often criticized by Triestino nationalists as too soft, too forgetful of the 'Slavic threat'. Thus, future developments of cross-border relations between Friuli-Venezia Giulia, Slovenia and Croatia will depend to a great extent on the internal balance between Friuli and Trieste.

Conclusion

There are perhaps two general lessons to be learnt from this account of the border-related experiences in the Alpe Adria area. The first concerns the presence of multiple factors – geographical, historical, military, political, cultural, economic and social. The historical factors should be placed both in the *longue durée* and in more short-term *évènementiel* history; the political factors should be analysed at different levels – international and inter-regional. All these levels and factors interplay in a complex fashion, which makes an orderly, consistent, theory-driven analysis very difficult. Complexity implies the intricacy, if not the impossibility, of forecasting the future. After 1989, social and political scientists have grown painfully aware of the limitations of their predictive abilities. I, at least, would be very hesitant in answering questions concerning the future of cross-border co-operation in the Alpe Adria region. And yet, an enduring faith in human rationality and goodwill makes me believe that co-operation will prosper, and that this area will become a model of transformation from a 'one-time genocide area' into an area of peaceful development, involving widely different ethnic and national groups.

The second lesson concerns the extreme difficulty of defining the concept of 'frontier region', and of assessing the role of borders and boundaries in affecting social, political and economic events in their vicinity. Almost all 'regions' in the Alpine-Danubian area – in Central Europe – are border regions, bounded by state frontiers; but the relevance of this factor seems to vary widely and there is no established social-science formula to measure this relevance. Borders affect not only the physical flow of goods and persons, which can be measured; much more important, they affect the culture and consciousness of people, which is much more difficult to assess. Moreover, they have functioned thus for centuries and even millennia, in different ways, and have left complex mental imprints.

The regions of Alpe Adria may have felt the need to build some form of common institutional arrangement because they physically touch each other, because they have common geographical borders, or because they have perceived common economic interests in interchanges and infra-structures, or because they felt the moral need to overcome ancient hatred, or because the memory of common membership in former political systems – for example, the Habsburg Empire – has prevailed, or because they share a common destiny within the new European Union. Or is it for some other reason? Or for all of the above? We do not know.

Many more technical questions concerning borders are raised by the Alpe Adria experience. One, for instance, has to do with the weakening or 'softening', or even 'withering away' or 'defunctionalization' of the internal frontiers of the European Union and its effects on the economy of the borderlands. Although certainly beneficial for the system as a whole, the weakening of frontiers may condemn border towns – whose main livelihood came from border controls and defence, and from the price difference in goods, wages, etc., on different sides of the frontier. Border economies are penalized by both extremes – complete opening and total closure of frontiers; they thrive when the differences and the degree of openness are 'just right'. Such adverse effects are already felt by some towns along the Austrian-Italian border, and compensatory measures are duly demanded.

Another problem derives from the fact that greater autonomy necessarily implies harder borders; when a political community dissolves in a plurality of sovereign states, new state boundaries arise. The hardening of the boundary between Slovenia and Croatia is a case in point. Although one may sym-pathize with the newly independent nations, the massive border structures built almost overnight between them cast a sombre shadow.

A final remark concerns the hiatus between political and economic integration and socio-cultural commonalty. The internal frontiers of Europe may well have been weakened to the point of disappearing, but the differ-ences in language, organization, attitudes, mores, values, information sources and outlook, remain important. The hardest frontiers are not those drawn on the ground, but those imprinted in the minds of people; and in this domain much remains to be done to bring about a real union of Europe.

Notes

1. See, for example, Regione Autonoma Trentino-Alto Adige (1981) *Regionalismus in Europa*, Munich: INTERREG.
2. See, for example, Regione Antonoma Trentino-Alto Adige (1992) 'Globalism and Localism: Theoretical Reflections and Some Evidence', in Z. Mlinar (ed.) *Globalisation and Territorial Identities*, Aldershot: Avebury; also A. Fedrigotti and G. Lerner (1993) *Alpe Adria: Identità e ruolo*, Trento: Regione Autonoma Trentino-Alto Adige.

3. See F. Gross (1978) *Ethics in a Borderland: An Inquiry into the Nature of Ethnicity and the Reduction of Tensions in a One-Time Genocide Area*, Westport, Conn.: Greenwood.

4. R. Strassoldo (1980) *From Barrier to Junction: Toward a Sociological Theory of Boundaries*, Gorizia: ISIG (mimeo).

5. C. Sambri (1970) *Una Frontiera Aperta: Indagini sui valichi Italo-Jugoslavi*, Bologna: Forni.

6. L. Buratti (1971) *La Frontiera Italiana: Introduzione e Testi*, Bologna: Forni.

7. G. Valussi (1972) *Il Confine Nord-Orientale d'Italia*, Trieste: Lint.

8. R. Gubert (1972) *La Situazione Confinaria*, Trieste: Lint.

9. R. Strassoldo (1972) *Sviluppo Regionale e Difesa Nazionale*, Trieste: Lint.

10. R. Strassoldo (ed.) (1973) *Confini e Regioni: Il Potenziale di Sviluppo e di Pace delle Periferie* (*Boundaries and Regions: Explorations in the Growth- and Peace-Potential of the Peripheries*), Trieste: Lint.

11. R. Strassoldo (1973) *Frontier Regions: An Analytical Study*, Strasbourg: Council of Europe (mimeo).

12. R. Strassoldo (1976) 'The Study of Boundaries: A Systems-Oriented, Multi-disciplinary, Bibliographical Essay', in *The Jerusalem Journal of International Relations*, **2**, 3; R. Strassoldo (1979) 'La Teoria del Confine', in R. Strassoldo (ed.) *Temi di Sociologia delle Relazioni Internazionali*, Gorizia: ISIG; R. Strassoldo (1981) 'Friuli-Venezia Giulia: a Border Region', in *Regionalismus in Europa*; and R. Strassoldo (1983) 'European Frontier Regions: Future Collaboration or Conflict', in M. Anderson (ed.) *Frontier Regions in Western Europe*, London: Frank Cass.

13. R. Strassoldo (1981); G. Delli Zotti (1983) *Relazioni Transnazionali e Cooperazione Transfrontaliera: Il Caso del Friuli-Venezia Giulia*, Milano: Angeli; G. Delli Zotti and B. De Marchi (1985) *Cooperazione Regionale nell'Area Alpina*, Milano: Angeli.

14. A. M. Boileau and E. Sussi (1981) *Dominanza e Minoranze: Immagini e Rapporti Interetnici al Confine Nord-Orientale*, Udine: Grillo.

15. R. Gubert (1976) *L'Identificazione Etnica: Indagine Sociologica in un'Area Plur-ilingue del Friuli-Venezia Giulia*, Udine: Del Bianco.

16. A. M. Boileau, R. Strassoldo and E. Sussi (1975) *Temi di Sociologica delle Relazioni Etniche*, Gorizia: ISIG.

17. F. Gross (1978).

18. R. Strassoldo and G. Delli Zotti (eds) (1981) *Co-operation and Conflict in Border Areas*, Milano: Angeli.

19. A. M. Boileau and B. De Marchi (eds) (1981) *Boundaries and Minorities in Western Europe*, Milano: Angeli.

20. See B. De Marchi (ed.) (1991) *La Communità Etnica Slovena Residente nelle Province di Gorizia e di Trieste*, Trieste: Regione Autonoma Friuli-Venezia Giulia; G. Delli Zotti and A. Rupel (1992) *Etnia e Sviluppo: Ruolo della Presenza Slovena nell'area Goriziana*, Gorizia: ISIG; and L. Bergnac and G. Delli Zotti (eds) (1994) *Etnie, Confini, Europa*, Milano: Angeli.

21. A detailed analysis of such activities can be found in Delli Zotti (1983).

22. See G. Delli Zotti (1994) *Dentro il Triangolo di Visegrad*, Gorizia: ISIG.

23. A. Sema (1994) 'Estate 1991: Gli amici Italiani di Lubiana', *Limes*, **1**.

24. See L. Bogliun-Debelju (1994) 'Come Faremo la Nostra Euregione Istria', *Limes*, **1**, pp. 263–70.

25. R. Strassoldo (1996) 'Ethnic Regionalism vs the State: The Case of Italy's Northern Leagues', in L. O'Dowd and T. M. Wilson (eds) *Borders, Nations, States*, Aldershot: Avebury.

Chapter 7

Mitteleuropa: The Difficult Frontier

EBERHARD BORT

Introduction: from 1989 to EU enlargement

After forty years of minimal change along what is now the eastern frontier of the European Union, we have witnessed, since 1989, a fundamental functional transformation of the old Iron Curtain, the tentative shaping of a new security architecture and the drawing of a new political map in Central Europe.

With the dismantling of the Iron Curtain, begun at the Austro-Hungarian border in the early summer of 1989, and all but completed with the fall of the Berlin Wall on 9 November of that pivotal year, an eastern frontier of the European Union, characterized by the dividing line through Germany and symbolized by this Berlin Wall, vanished. The frontiers between the European Union and its eastern neighbours have been opened. With the integration of the former German Democratic Republic into Germany, the frontier was moved eastward. With the envisaged accession of Poland, the Czech Republic and Hungary into the European Union, 'central and eastern Europe is becoming part of the West. Yet,' David Marsh also observed,

> the economic and social divisions between East and West Germany provide a reflection and a reminder of the large disparities in the organisation of lives, economies and states throughout Europe. Germany is the focus and the mirror of a continent united in its disunity.[1]

A 'snapshot' of the Polish-German border illustrates the nature of the change from an 'alienated', sealed frontier, to an open border:[2]

- Border crossings increased from 1990 to 1991 by 40 per cent; cross-border crime in the same period by 400 per cent.
- 297 million people crossed the Polish borders in 1995.
- Between 1991 and 1995, 48,000 people were arrested at the border, 35,000 of them illegal immigrants.
- In 1995, 1319 false passports were confiscated, 400 false visas and 500 other forged travel documents.

At the German-Czech border, the data spell out a similar message:[3]

- In 1991, 59 million persons crossed the German-Czech frontier; in 1995 the figure had increased to 98 million.
- In 1991, 17 million cars were counted at the German-Czech border; in 1995, that figure had reached 30 million.
- In 1991, 18,000 persons were held up trying to cross the border illegally from east to west; in 1993, this figure rose to 43,000 persons. Due to very efficient additional border controls after the 'velvet divorce' of Czechoslovakia at the Czech-Slovak frontier, the figure for 1995 was down again to 19,000 persons.

New checkpoints had to be installed, and the number of personnel at the border drastically increased to cope with the dramatic rise in cross-border traffic and cross-border trade. Despite these efforts, supported by PHARE money from Brussels, waiting times of over a dozen hours at Polish or Czech border crossings are still a common and, for the export and import of goods, an expensive experience.

All of the Central and Eastern European Countries (CEEC) have now applied for EU membership, a membership which, in principle, is accepted by the EU. Enlargement seems no longer a question of 'if', but of 'when' and 'whom'.[4] The internal legal reforms and the transformation into fully working market economies in the CEEC are outlined in the 'Europe Agreements' signed with all potential new EU Member States in the early 1990s.[5]

Applicants, Jacques Santer said on 19 January 1996 in Germany, must become fit for accession, the Union must become fit for expansion – and he called the latter the most important task of the Intergovernmental Conference (IGC).

The IGC, started in March 1996, due to terminate in 1997, faces the task of preparing the internal structure of the Union of the Fifteen to cope with enlargement, which could bring the membership of the EU up to between 25 and 30 within the next decade. 'Institutional procedures introduced to deal with six Member States', David Martin has succinctly argued, 'are coping badly with 15; reform is needed if the Union is to function with the proposed 27 members next century.'[6] Moreover, without radical reform, or long phasing-in periods under maintenance of border controls and customs, enlargement would, in terms of the EU's Common Agricultural Policy, simply not be manageable.[7] The same could be said for structural funds and the Cohesion Fund.

The outlook, though, seems far from promising. It is generally assumed that the 'deepening' of the decision-making structures within the European institutions will not go beyond superficial cosmetics (reform in regional funding or the Common Agricultural Policy are outside the remit of the IGC), but that enlargement will come nonetheless. Pressure of an enlarged EU may lead to reforms not possible under the present circumstances, but there is also the fear of a break-up of the Union into a 'core Europe' and peripheral, 'second-rate' member states, illustrated by the process of European Monetary Union, or even a regression of the Union to the level of a Free Trade Zone.[8] A multi-speed Europe, as envisaged by the joint initiative of France and Germany of October 1996, seems inevitable.

Overlapping with the project of EU enlargement is the planned expansion of NATO, which will create another set of 'ins' and 'outs' in Central Europe, marking a new security frontier across the continent.

A new political map in Central Europe

Central Eastern Europe is, in the words of William Wallace, a 'region faced with complex and difficult demands for political adaptation'.[9] A whole mosaic of different ethnic, linguistic and religious groups, suppressed under the conditions of the 'Cold War' and Communist rule, has emerged, revealing deeply rooted nationalisms, marked by the re-surfacing of ancient quarrels and historic disputes as well as tensions between what could be termed urban-liberal attitudes towards the European Union and rural communities clinging to traditional, conservative values, mostly out of fear of the possible consequences of joining the EU.[10]

There are, to single out but a few of the conflicts which seemed buried under the Cold War conditions, the dissent between Hungary and Slovakia over the Gabcikovo-Nagymaros dam system on the Danube,[11] the status of the Hungarian minority in Slovakia,[12] and Italian-Slovenian border frictions over the status of the Italian minority in Slovenia.[13] German–Czech differences over the Sudeten question were not fully resolved by the accord of January 1997.[14]

Before the fall of the Wall, the 1980s had given renewed rise to the idea of a common identity in *Mitteleuropa*, in the sense of Milan Kundera's famous phrase: '*Mitteleuropa* is not a state, but a culture, a fate.'[15] This was coined when intellectuals on both sides of the Iron Curtain felt the threat of their borderlands becoming a potential theatre of a 'winnable' nuclear conflict between the superpowers, with the stationing of Pershing and Cruise missiles on one and SS20s on the other side of the divide.

Kundera's statement, and the equally important contribution of György Konrád from Hungary,[16] date mainly from the mid-1980s and are inextricably tied up with the experience of the Iron Curtain as an unnatural dividing line in the centre of the continent, particularly after Helsinki 1975. The overpowering threat of the 'other' in East and West encouraged the peoples

Table 7.1 Labour productivity in manufacturing (output per employee), percentage change

	1992	1993	1994	1995
Bulgaria	0.2	5.5	14.2	10.7
Czech Republic	−7.6	−3.5	4.0	18.4
Hungary	10.6	18.4	7.3	13.0
Poland	17.1	14.5	19.2	12.3
Romania	−10.0	10.4	11.6	19.2
Slovak Republic	7.4	0.6	6.8	4.3
Russia		−14.2	−13.7	2.9
Germany	10.5	−2.2	−4.7	13.6
United Kingdom	1.8	−14.6	2.0	6.7

Source European Bank for Reconstruction and Management

of *Mitteleuropa* to rediscover a sense of a shared and endangered cultural heritage, as a 'family of cultures',[17] subsumed under the label 'European' or 'Central European'. 'East-Central Europeans' are, however, not harking back to Naumann's concept of *Mitteleuropa*,[18] which has – unfairly, perhaps – been described as a pan-German imperial construct, but looking forward to federal and regional structures as the only adequate form of state organization in this area:

> In this area, the homogenous nation state is the exception; as a norm it is useless. . . . The central European idea means the flourishing plurality of its constituent parts, the consciousness of diversity.[19]

Compared to pre-1989, the concept of the 'Cold War', the 'concept of "West" and "East" lose[s its] common *principium divisoris*'.[20] The post-1989 period is, however, characterized by paradoxical developments. The West plunged into a deep recession just at the time the borders to the East opened; in stark contrast, some eastern neighbours of the EU started, as the shock-waves of the 'velvet revolutions' ebbed, to perform economically so successfully that they have been compared to the Pacific 'Tiger economies'.

Just as the Iron Curtain was lifted, bringing in its wake a softening and opening of frontiers, this border became, under the auspices of 'Europe '92' and Schengen, the external frontier of the European Union, which had to be hardened; and controls along it had to be strengthened to allow for the dismantling of the EU's internal frontiers. But – yet another paradox – policing this border has entailed the creation of a border zone where random checks are allowed.[21] Dr Horst Eisel, Assistant Director for Frontiers at the German Ministry of Internal Affairs, has put it thus:

> The spatial approach clearly ought to take precedence over the purely linear approach to geographic boundaries. The latter is no longer a match for today's challenges, because individual and collective security begins beyond our borders and continues well on this side of them.[22]

Germany's Minister of the Interior, Manfred Kanther, a law-and-order hardliner within the ruling CDU party, has over the past few years targeted this border zone which, in his words, has become a 'crime zone'. Apart from installing state-of-the-art surveillance technology he had, by May 1996, brought up the number of border police (*Bundesgrenzschutz*) at Germany's eastern frontier to 5500; in January 1997, he announced an extra 1500 *Grenzschützer* – which means that more border police are stationed at Germany's eastern borders than the number of illegal immigrants counted there in 1996 (5000 of the estimated 7000 who tried to cross the borders were detained; in 1995 the figure was 4000). As the *Süddeutsche Zeitung* noted, there is a link between the new, restrictive asylum laws of Germany and the rise in illegal immigration: by closing the doors on asylum, foreign refugees are now driven into the arms of organized human traffickers, paying up to $5000 a head for their criminal services.[23] As the Schleswig-Holstein Minister of the Interior, Ekkehard Wienholtz, explained at a Danish–German meeting in Flensburg, Kanther's reinforcements at the German–Polish and German–Czech borders have already had an effect: more illegal migration and smuggling is now channelled through the Baltic states and Scandinavia.[24]

Although it might well be argued that the drive towards integration in the West has, as it were, suffered from the eclipse of a uniting ideological enemy in the East, there is still a general tendency towards integration, region-alization, even federalization within the EU. However, in many of the newly independent states to the east of the EU strong tendencies towards con-solidating the regained sovereignty of the nation-state prevail, having supplanted at least part of the notion of a common *Mitteleuropa*. In many of these countries, the contradiction between emphasis on national sovereignty and the aspiration to join the European Union is not openly discussed, leaving the field to unreconstructed Communists and the nationalist right, mostly still on the fringes, but ready to exploit any disillusionment with European integration.

An alternative approach may be illustrated by the case of Ireland. The former Taoiseach (Prime Minister) of Ireland, Garret FitzGerald, and the political scientist Paul Gillespie analysed that 'EU membership has completed the project of Irish independence'.[25] It compensated for the dominant British economic and political influence on Ireland and 'introduced into the Anglo-Irish relationship both greater equality and a more neutral arena in which to conduct relations'. Moreover, pooling sovereignty with its partners in the EU, they argue, has 'released the country from the inward-looking nationalism of the four decades after independence'.[26] The Irish ambassador in London, Ted Barrington, explained to the *Observer*:

There's a psychological dimension as well as an economic one. Our whole experience of membership is that it enhances our sovereignty in a real sense, if you define sovereignty as a capacity to define your own destiny, by being able to take decisions on huge issues which affect the economic welfare of the continent.[27]

The CEEC on the threshold of entering the EU, having stepped out of the shadow of the Russian empire, seems to view membership primarily as a security shield and the opportunity to gain the same living standards as the West. The question of pooling sovereignty is eschewed, as when the Czech Prime Minister, Vaclav Klaus, declared that a referendum on EU membership was only necessary if national sovereignty would have to be ceded to EU institutions,[28] as if the *acquis communautaire*, acceptance of which is the basis of any EU membership,[29] would not clearly demand just this. Refusing to explain the issues openly to the electorate – and there might be parallels with Britain's present 'Eurosceptic' problems – could prepare the ground for long-term difficulties in the integration process. Looking, on the other hand, at Ireland's example (or, for that matter, Scotland's aspiration towards a greater degree of self-government[30]) might show how the voluntary sharing of sovereignty – in contrast to involuntarily being a colony or a vassal – with a real voice in how your own and European affairs are being run, can have a liberating effect, and leave less space for ultra-nationalist demagogy.

This is made even more complex by the fact that, at least in the case of Poland, the Czech Republic and Hungary, aspirations to membership in the EU are intertwined with their interest in NATO enlargement. President Clinton, in a significant speech in Detroit on 22 October 1996, declared NATO enlargement to be on the cards, envisaging the acceptance of new members by 1999. Although he did not mention any specific states, it was widely held that he referred to Poland, the Czech Republic and Hungary (see Figure 7.1). This would mean that a new European security frontier would cut through *Mitteleuropa*.

The immediate reaction by former Soviet president Mikhail Gorbachev was 'a chilling warning that a new Iron Curtain could come crashing down in Europe, reviving old fears and the threat of nuclear conflict',[31] urging leaders in the West to rethink their NATO strategy: 'If they do not, there will be a new division, new suspicion, a new Iron Curtain and arms race.'[32]

What will be the consequences for those not deemed eligible for NATO expansion? This first round of expansion will, if it happens as expected, include those states arguably least in need of the NATO umbrella. Will there be security guarantees for the states further east? Or are they destined to become a grey zone, a buffer area between Russia and the West? Or will they even fall back under Russian influence? Growing bitterness is apparent in the

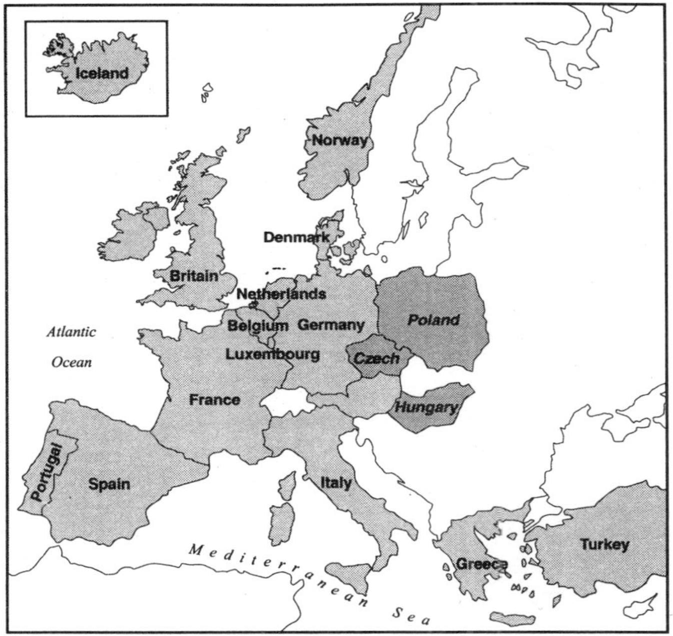

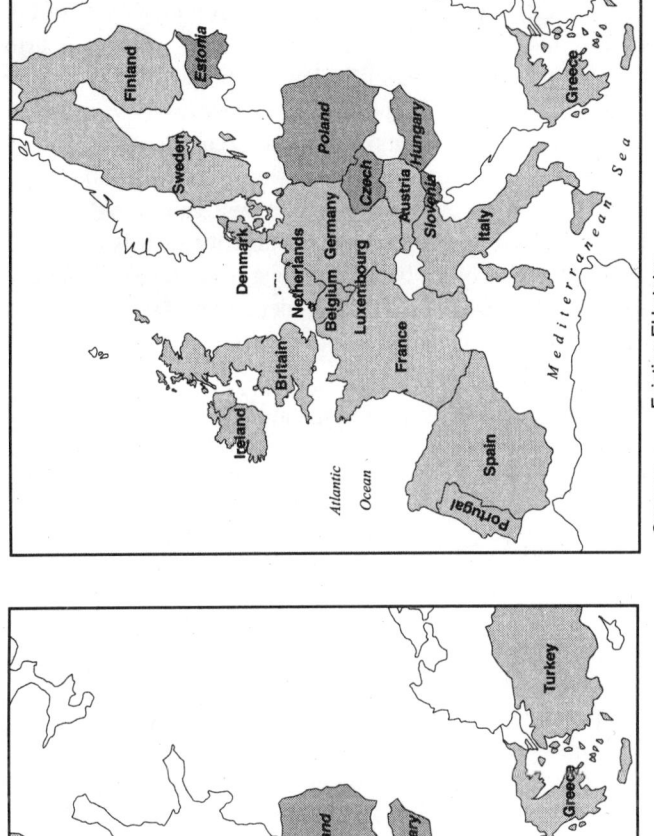

Germany   Existing NATO states

Hungary   Proposed NATO enlargement

Figure 7.1   NATO enlargement

Germany   Existing EU states

Hungary   Proposed EU enlargement

Figure 7.2   EU enlargement

Baltic States about lack of support from the West against permanent Russian pressure.[33]

The German reaction to President Clinton's re-election was illuminating in this context. Karl Lamers, the governing CDU foreign policy spokesman, reminded Clinton that his NATO expansion project is in need of a firm agreement with Russia, pointing out that 'Nobody is helped if the security situation is more tense after a NATO expansion than before.'[34]

Likewise, EU enlargement will have 'ins' and 'outs', creating a new eastern frontier of an enlarged European Union, in all likelihood not congruous with the new NATO frontier line. What consequences will that have for the development of both a regional and a European identity in this area and beyond? These new border lines could become dangerous for the workings of an enlarged European Union if potential cross-border conflicts straddling this new frontier are not fully and satisfactorily settled by the time enlargement happens:

> Such conflicts are likely to be very divisive in the internal politics of the EU because the member states and important interests will diverge sharply on the policies to be adopted. The divisions caused could threaten the cohesion of the Union.[35]

Lastly, the Turkish Foreign Minister, Tansu Çiller, has threatened to veto NATO enlargement if Turkey is not allowed into the European Union. What looks like blackmail is in fact disappointment at the EU withholding (on the behest of Greece) subsidies agreed on when the customs union between the EU and Turkey came into effect in January 1996; moreover, it is the desperate cry for support of a secular, Western-oriented Turkey allegedly on the brink of being swallowed by Islamic fundamentalism. Withholding EU support could mean playing into the hands of the Islamic fundamentalist leader and present Prime Minister Erbakan and help his unhindered march through the institutions.[36]

Cross-border co-operation

The successful development of institutionalized cross-border co-operation at local and regional levels in Western Europe is part of the integrationist project, but it also transcends the frontiers of the European Union. It has been seen as an important project which should be followed along the eastern frontier of the European Union since the pivotal changes of 1989/90.[37] Transfrontier co-operation is interpreted as a cornerstone of European integration, because its aim is to transform traditionally peripheral border-lands into prosperous core regions of Europe. In the terminology of Oscar Martinez, the internal frontiers of the European Union would be classified as 'integrated borderlands', contrasting with the areas adjacent to the former

Iron Curtain, at the other end of the spectrum, which was an example of 'alienated borderlands'.[38]

Frontiers, historical dividing lines, separating rather than linking, have been transformed into junctions of regions or states; frontiers, historically often disputed, and the focus of territorial conflicts, have turned into borderlands with intensified exchanges and a high degree of interdependence, promoting peaceful development and co-operation. The European Parliament has repeatedly stressed the significance of border regions for the construction of the European Union 'in all its dimensions', but particularly in developing the relationship with the new democracies in Eastern Europe.[39]

Wherever possible, the newly established Euregios along the EU's eastern frontier build on local and regional initiatives. The people living and working in these regions are also sharing a common experience as having been peripheral in their national contexts, with cul-de-sac transport systems and a frontier mentality developed over forty years living along the Iron Curtain. The eastern Euregios are also clearly modelled on their western forebears. Their aims can be described as targeting three main areas:[40]

1. *Economic co-operation.* Euregios aim at strengthening the economy of the cross-border region, by co-ordinating infrastructure planning, optimal use of resources both sides of the border and complementary investment (avoiding unnecessary duplications). Their first objectives are creating more, and more efficient, border crossings, and improved communications between economic players both sides of the border. They also stress technology transfer and easing border crossings for transfrontier workers.

2. *Ecologically based, sustainable tourism.* Ecological projects, from recycling to tackling sewage and refuse problems, are high on the Euregio agendas. Common cross-border landscape protection projects have been initiated to prevent modernization and industrialization destroying the one advantage of having been a peripheral zone – that is, untouched areas of natural beauty. The aim is to preserve them for environmentally friendly, sustainable tourism – from the lakes of Pomerania, through the mountain areas along the German–Czech border, down to the Danube national park (finally agreed in October 1996[41]).

3. *Culture and communication.* Another main goal of Euregio activity is the creation of – or support of – a regional consciousness, a regional identity, whether by highlighting a common historical heritage (the name 'Pomerania' harks back to the Latin name of the region), or common industrial and artisan traditions (for example the 'Bohemian' glass industry[42] in the Czech, Bavarian and Austrian borderlands of the Regio

Bayerischer Wald/Böhmerwald), or by fostering cultural exchange and, in particular, youth exchange, educational co-operation of schools and universities, increased contacts and exchange between local authorities, social institutions and sports clubs.

Euregio organizations, furthermore, co-ordinate infrastructure and spatial planning to apply for EU financial support under the INTERREG and PHARE programmes, and to use efficiently these programmes. Both programmes, it could be argued, aim at providing an 'as if' situation along the European Union's eastern frontier – as if the adjoining borderlands in the East were already part of the EU. The question must be asked, what consequences this might have for the development of regional, national and European identities.

It seems that the establishment of cross-border co-operation – Euregios from Finland to Slovenia – shows that a translation from west to east is well under way. The felt need for cross-border co-operation (environment, infrastructure, tourism, security) matches certain regional reform concepts, devolving planning authority and decision-making processes to the regions. The regional context may also be more conducive to solving problems of national minorities and even provide regional solutions for international problems. Cross-border regionalism seemed to develop, for instance, in the German–Czech borderlands despite the fraught negotiations between Germany and the Czech Republic about coming to a final agreement about the property of, and compensation for, the Sudeten Germans. The accord of January 1997, when the Czech Premier Vaclav Klaus and the German Chancellor Helmut Kohl signed a post-war pact aimed at ending half a century of resentment and bitterness between the two neighbouring countries, may open up new opportunities for intensified cross-border co-operation.[43]

There are, however, reasons for being cautious about a simple transfer of Western models to the Eastern borderlands. Different centre–periphery relations ought to be considered. Jutta Seidel of the State Chancellery of the Free State of Saxony, speaking about her experience in organizing transfrontier co-operation between Saxony and Poland, and Saxony and the Czech Republic, emphasized the initial difficulties to be overcome, resulting from administrative centralization in the Czech and Polish Republics. According to Frau Seidel, it took years to create trust and willingness in Warsaw and Prague to allow the border regions a degree of planning autonomy, matching that of their German counterparts.[44] This evokes a general problem of federal states (Germany, Austria) on one side of the frontier, and centralized nation-states on the other. Regionalization and 'integrated borderlands', rather than a nineteenth-century model of the nation-state could, however, offer a more tranquil future for non-homogenous states with large ethnic minorities within their borders.

The potential of conflict between centre and periphery may be exacerbated by the major difference between the West and the East. While in the West, especially along the German frontier, states with roughly similar wage levels and costs of living started collaborating across their borders, the eastern frontier of the European Union has become, under the conditions of the Cold War, a profound economic frontier. If, as Malcolm Anderson has argued, 'for the countries bordering the EU to the east, problems of extreme economic inequality have caused anxieties about economic and political subordination to them',[45] could not other regions within Poland, the Czech Republic and other Central and Eastern European countries also ask the question: why should the regions bordering on the EU be privileged? In other words, is the new, perhaps common, identity of the transfrontier regions a forerunner for integrating the whole of the states concerned into the EU, or does it cause resentment against these apparently privileged regions? Does it reinforce the considerable disparities between East and West, town and country, within the applicant states?[46] The question could also be put the other way: is it possible to stop the eastern borderlands gaining advantages without closing the border?

What kind of new identity is actually being created in *Mitteleuropa*? This is a question not only asked in the area concerned, but also in the Western part of Europe, in the light of the future role of Germany. Would any revival of a notion of *Mitteleuropa*, including Germany, perhaps even dominated by Germany, put a new dividing line between Western Europe and this *Mitteleuropa*? And would this dividing line separate Germany and France, who so far have seen their common frontier as the axis of European integration? Asked the other way round: is any notion of *Mitteleuropa*, containing Germany, at all possible as long as France and Germany see themselves as the core of Europe?

The Western partners along the eastern frontier of the EU support cross-border co-operation primarily, no doubt, to help their own borderlands, which severely crippled their economic development by having been in the shadow of the Iron Curtain (cf. the German '*Zonenrandgebiet*'). As far as the official policy of the European Union is concerned, the founding of Euregios along the eastern frontier is certainly seen as supporting the claim of the applicant states *in toto* to become, in the words of EU President Jacques Santer, 'fit for accession'. Yet there is distrust of the Euregios in the eastern capitals. The Visegrád states have been very careful to define their relationships so as not to create any impression that closer transfrontier co-operation in *Mitteleuropa* could be understood as a 'Half-Way House' on a path towards full membership of the European Union.[47] Cross-border regionalism, in their view, must not be a substitute for the fastest possible track into the EU for the applicant state. Likewise, they warn against any kind of 'bloc mentality' in the West with regard to the individual applications of Central and Eastern European states. In the case of Hungary, András Inotai has argued that:

> in various areas, as infrastructure or protection of the environment, the regional approach of the EU would be welcome; however, this must not lead to the revival of 'bloc mentality', or to the identical treatment of different associated countries, as they are on rather different levels of political, social and economic development.[48]

And he continued:

> as widespread international experience shows, the meaningful extension of regional co-operation is not a precondition for, but a consequence of EU membership.[49]

Conclusion: new European identity or 'clash of civilizations'?

'Expanding the Union is a potentially divisive issue', wrote Sarah Helm in *The Independent*, 'as it will require radical reforms of agricultural and regional aid spending, potentially draining EU funds. An expanded EU will also increase the need for streamlined decision-making within its institutions'.[50] Alois Berger, in the Berlin *tageszeitung*, pointed out that the EU may become totally paralysed: 'Without a restriction on the veto as it now exists decision making in a EU of 18 or more is not possible.'[51] And Mark Frankland, in the *Observer*, remarked gloomily that 'the EU, preoccupied with internal problems, seems oblivious to the danger of leaving East Europe too long outside its gates. It has no realistic plan to gather in the Poles and Czechs before the millennium is up.'[52]

If the problem is not seriously addressed, and if there is no adequate solution by the middle of 1997 when the IGC is supposed to close, this must be taken as a serious set-back to the membership aspirations of the applicant states. Will a paralysed European Union really be capable of offering enlargement? Will a paralysed European Union be worth joining? Can the ability of an enlarged EU to respond to increased reform pressures be trusted? Will popular backing for membership in the applicant countries collapse after an unsatisfactory outcome of the IGC?

There is a strange contradiction between the disenchantment with the post-Maastricht European Union within the Member States (as the narrow referenda results in Denmark and France showed, and the outcome of the recent Euro-elections in Austria and Finland) and the eagerness of the Central and Eastern European countries to become members of the club. Figures for Hungary, however, seem to show a certain erosion of popular backing for EU membership. In answer to the question: 'What is your opinion on the aims and activities of the European Union?', the 'positive' figure fell – between 1991 and 1993 – from 45.2 per cent to 35.4 per cent; the 'neutral' response rose from 28 per cent to 32.1 per cent; the 'negative' stance increased from 4.3 per cent to 8.3 per cent.[53] Norway and Switzerland have chosen to stay outside the EU, and the recent Labour victory in Malta can be seen as the de facto withdrawal of Malta's 1990 EU application.

Despite these difficulties, transfrontier regional co-operation makes progress, from the Pomerania in the North to the Region Triagonale between Austria, Hungary and Slovakia. Broad agreement exists that cross-border co-operation is desirable, that its institutionalization creates good neighbours and removes potential sources of international conflict, and is thus conducive to peaceful development, prosperity and stability. Tony Asiwaju's observation is probably justified that cross-border co-operation as witnessed in the West provides the only viable model for their Eastern counterparts.[54] But potential rivalries and conflicts between local, regional, national and supranational levels of co-operation must not be ignored. At best, these levels complement each other, creating a European identity in diversity but, at worst, they could disrupt the European project.

The current situation in Western Europe has been the product of over 40 years of collaborative efforts between roughly equal partners, with less difficult linguistic, cultural and economic boundaries than those to the east of the old Iron Curtain. The CEEC have only recently regained their independence, their right to self-determination, their sovereignty, their ability to formulate freely their national identity. Pooling these precious gains with their Western neighbours is a process which needs time. Borders, as Michael Bosch concluded at a recent conference on the eastern frontier of the European Union, are 'also seen by some CEEC as a protection against an all-too fast assimilation of their young republics'.[55]

When, in November 1996, newspapers reported that Austria would delay the implementation of the Schengen accord, in the words of Austria's then Minister of the Interior, Caspar Einem, 'due to inadequacies in the Schengen Information System',[56] this piece of news was received with relief in Slovenia. It would take time to adjust to border controls, and the more time, the better.

But cross-border regionalism seems the best way forward for some purposes. It can tackle infrastructure problems directly, and it can encourage a bottom-up approach to building a more democratic Europe. As indicated above, the problems of adaptation of the countries of East Central Europe necessitate long transitional periods until fully integrated EU membership can be achieved. Functioning cross-border arrangements can make this transition smoother and may well accelerate the process.

Transfrontier regionalism may also be a means of 'discovering that "family of cultures" . . . through which over several generations some loose, over-arching political identity and community might gradually be forged'.[57] It is too early to claim that this process is already 'over the hump'. It is a complex process and will take time:

> Shaping a cultural identity that will be both distinctive and inclusive, differentiating yet assimilative, may yet constitute the supreme challenge for a Europe that seeks to create itself out of its ancient family of ethnic cultures.[58]

The positive, optimistic perspective is that Europe will meet this challenge. The recent election results in former Yugoslavia, Bulgaria and Romania, as well as the increasing protest movement in these countries, have been seen in the West as bringing these states 'closer to Europe',[59] enhancing the process of 'get[ting] rid of the East European political identity',[60] as Anton Bebler called it when he described how Slovenia 'dissociated herself from the Balkan crisis'.[61] This could be marking a turning point in the post-1989 process of creating potential 'ins' and 'outs' for the various schemes of European architecture to the east of the existing European Union.

But there are less hopeful, more ominous perspectives as well. Despite Germany's massive financial transfers to former East Germany, the dividing line along the former Iron Curtain (and through Berlin) is still there, not as a visible, tangible border, but in the difference of living standards and as a 'border of the mind', as successive elections have shown in the years since unification.[62] The difficulties in integrating the relatively privileged former GDR into the strongest economy of the West cast a negative shadow on the eastern enlargement project of the EU.

The emergence of a new security architecture in Europe will probably create new frontiers, perhaps, as some commentators envisage, along the lines of Samuel Huntington's 'clash of civilizations', with those countries who are ethnically Slav and religiously Orthodox or Islamic as the natural allies of Russia, whereas those religiously Christian (Catholic or Protestant) and linked in the past to the Holy Roman and Habsburg Empires aligning with the West.[63]

Jean-Christophe Rufin sees the East–West divide already replaced by the 'new *limes*', the North–South divide running through the Mediterranean, but this may be a slightly premature prediction.[64] The eastern frontier of the European Union, even when relocated further east, will remain – in the American usage of the term – a 'difficult frontier'; it will be the terrain, in Perry Anderson's words, of 'political quicksands on which the Europe to come will be built'.[65]

Notes

1. David Marsh (1995) *Germany and Europe: The Crisis of Unity*, London: Heine-mann [1994], Mandarin, 1995, p. 174. See also H. Buchholz (1994) 'The Inner-German Border: Consequences of its Establishment and Abolition', in C. Grundy-Watt (ed.) *Eurasia* (World Boundaries Series), London: Routledge, pp. 55–62.

2. Data provided by the Director of the Polish border post of Olszyna/Forst, Marian Kalek, the Chief of Border Police, Colonel Jerzy Piwowarski, and the Deputy General Commandant of the Polish Border Police, Colonel Wojciech Brochwicz; see E. Bort (1996) 'A View from the Polish Border', in M. Anderson and E. Bort (eds) *Boundaries and Identities: The Eastern Frontier of the European Union*, Edin-burgh: ISSI, pp. 57–60.

3. Data provided by Ivo Schwarz, Director of Foreign and Border Police, Plzen, and Dr Milos Mrkvica, Director of the Department of Migration/Foreign and Border Police, Prague; see E. Bort (1996) 'Coping with a New Situation', in M. Anderson and E. Bort (eds) *Boundaries and Identities: The Eastern Frontier of the European Union*, Edinburgh: ISSI, pp. 61–2.

4. See Heinz Koehler, 'Die Perspektiven der Erweiterung der EU um die mittel- und osteuropäischen Länder (MOEL)', in M. Anderson and E. Bort (eds) *Boundaries and Identities: The Eastern Frontier of the European Union*, Edinburgh: ISSI, pp. 81–9.

5. See Peter-Christian Müller-Graf (1993 edition) *East Central European States and the European Communities: Legal Adaptation to the Market Economy*, Baden-Baden: Nomos; see also ECSA (1993 edition) *The Legal, Economic and Administrative Adaptations of Central European Countries to the European Community*, Baden-Baden: Nomos.

6. David Martin MEP (1995) *1996 and All That*, Dalkeith: Group of the Party of European Socialists, p. 5.

7. In the Central and Eastern European states 25 per cent of the country's workforce are employed in agriculture; in the EU, the figure is a mere 5.7 per cent! (see 'MOE-Staaten auf EU-Kurs', *Die Welt*, 2 January 1997).

8. See Heinz Koehler (1996), *op. cit.*

9. William Wallace (1994) 'Rescue or Retreat?', *Political Studies*, **42**, p. 76.

10. For a short historical overview of the development of frontiers in Central and Eastern Europe, see M. Anderson (1996) *Frontiers: Territory and State Formation in the Modern World*, Cambridge: Polity Press, pp. 56–66.

11. See John Fitzmaurice (1996) *Damming the Danube: Gabcikovo and Post-Communist Politics in Europe*, Boulder, Col.: Westview.

12. See Werner Weidenfeld and Manfred Huterer (1992) *Osteuropa: Herausforder-ungen – Probleme – Strategien*, Gütersloh: Bertelsmann, pp. 24–6.

13. Anton Bebler (1994) 'The Republic of Slovenia – Internal Politics and Inter-national Relations', *Perspectives*, **4** (Winter 1994/5), Prague: Institute of International Relations, pp. 25–33, in particular pp. 31–2.

14. See, for example, Alexander Gorkow, 'Bayerisches Störfeuer nach Bonn und Prag', *Süddeutsche Zeitung*, 24 October 1996; and 'Kohl stiftet in Prag neue Verwirrung', *Süddeutsche Zeitung*, 23 January 1997. For the background, see Bundeszentrale für Politische Bildung (1995 edition), *Tschechen, Slowaken und Deutsche: Nachbarn in Europa*, Bonn.

15. Milan Kundera (1984) 'A Kidnapped West or Culture Bows Out', *Granta*, **11**, p. 106.
16. György Konrád (1985) *Antipolitik: Mitteleuropäische Meditationen*, Frankfurt: Suhrkamp.
17. See A. Smith (1992) 'National Identities and the Idea of European Unity', *International Affairs*, **68**, especially pp. 74–6; excerpts reprinted in M. O'Neill (1995) *The Politics of European Integration: A Reader*, London: Routledge, pp. 314–19.
18. See also Jiři Gruša (1996) '"Ich will die Grenze loben": Gedanken zur Kultur der Grenze', in M. Anderson and E. Bort (eds) *Boundaries and Identities: The Eastern Frontier of the European Union*, Edinburgh: ISSI, pp. 27–38, 31.
19. György Konrád (1986) 'Der Traum von Mitteleuropa', in E. Busek and G. Wilflinger (eds) *Aufbruch nach Mitteleuropa: Rekonstruktion eines versunkenen Kontinents*, Wien: Edition Atelier, pp. 87–98, p. 92 (my translation). For a historical discussion of the development of the *Mitteleuropa* idea, see also M. Schubert (1993) *Die Mitteleuropa-Konzeption Friedrich Naumanns und die Mitteleuropa-Debatte der 80er Jahre*, Sindelfingen: Libertas.
20. Miroslav Kusy (1989) 'We, Central-Europeans–Western-Europeans', in G. Schopflin and N. Wood (eds) *In Search of Central Europe*, Cambridge: Polity Press, p. 91.
21. Generally speaking, Schengen envisages a 20 km border zone, but in reality, for instance, the whole of Bavaria is now defined as a 'border zone'.
22. Quoted in Patrice Molle (1996) *External Borders Pilot Project: Placement Report*, Strasbourg: Centre des Etudes Européennes, p. 6.
23. See Peter Scherer (1996) 'Kanther verstärkt Sicherung der Ostgrenzen', *Die Welt*, 8 May; and Christoph Schwennicke (1997) 'Abwehrmauer an den Ost-grenzen', *Süddeutsche Zeitung*, 3 January.
24. Diethart Goos (1997) 'Skepsis über çchengener Vertrag', *Die Welt*, 11 February.
25. Garret FitzGerald and Paul Gillespie (1996) 'Ireland's British Question', *Prospect*, (October), pp. 22–6.
26. Ibid.
27. Quoted in Arnold Kemp (1997) 'Which country is enjoying a European economic boom, Germany or Ireland? Take your time . . . ', *The Observer*, 2 February.
28. See Michael Frank (1996) 'Die EU möge sich bitte nicht zuviel herausnehmen', *Süddeutsche Zeitung*, 27 March.
29. According to the criteria for new members laid down at the Copenhagen Summit, June 1993.
30. John Palmer, the European editor of *The Guardian*, spelled out the British context in the *New Statesman*: 'Indeed, devolution, plus proportional representation, plus a Bill of Rights, together with steps to a more democratic and federalising European Union, pose the biggest threat to the unaccountable, secretive, obso-lete, undemocratic and corrupt UK state since the great pro-reform mass movements in the 19th century.' John Palmer (1996) 'The Future is Federal', *New Statesman*, 22 March, pp. 16–17.
31. Herbert Pearson (1996) 'Gorbachev warns of new Iron Curtain', *The European*, 31 October, pp. 1–2.

32. Ibid.
33. See Carl Gustav Ströhm (1997) 'An der Ostsee wächst die Verbitterung über den Westen', *Die Welt*, 18 January.
34. A Reuters report quoted in *The Irish Times*, 6 November 1996.
35. Malcolm Anderson (1996), *Frontiers*, p. 181.
36. Josef Joffe (1997) 'Die rostige Pistole der Frau Çiller', *Süddeutsche Zeitung*, 30 January; and Evangolos Antonaros (1997) 'Ankaras Forderung an die EU ist ein politischer Hilferuf', *Die Welt*, 31 January.
37. See A. L. Asiwaju (1996) 'Public Policy for Overcoming Marginalisation: Borderlands in Africa, North America and Western Europe', in S. Nolutshungu (ed.) *Margins of Insecurity: Minorities and International Security*, Rochester, NY: Rochester University Press, pp. 251–84, see particularly pp. 282–4. See also Hans Briner (1996) 'Regio Basiliensis als Modell: Das Europa der Regionen – die Perspektive des 21. Jahrhunderts', in M. Anderson and E. Bort (eds) *Boundaries and Identities: The Eastern Frontier of the European Union*, Edinburgh: ISSI, pp. 39–46. The fundamental importance of transfrontier co-operation for European integration has been expressed in Victor von Malchus (1975) *Partnerschaft an europäischen Grenzen: Integration durch grenzüberschreitende Zusammenarbeit*, Bonn: Europa Union Verlag. A more recent overview and analysis can be found in Silvia Raich (1995) *Grenzüberschreitende Zusammenarbeit in einem 'Europa der Regionen'*, Baden-Baden: Nomos.
38. O. J. Martinez (1994) 'The Dynamics of Border Integration: New Approaches to Border Analysis', in C. H. Schofield (ed.) *Global Boundaries*, London: Routledge, pp. 1–15.
39. John Cushnahan (1992) *Bericht des Ausschusses für Regionalpolitik, Raumordnung und Beziehungen zu den regionalen und lokalen Körperschaften über die grenzüberschreitende und interregionale Zusammenarbeit*, European Parliament, 21 May, p. 5.
40. Cf. Jens Schwarz (1995) *Perspektiven grenzüberschreitender Tätigkeit im Rahmen der Europaregion POMERANIA*, Sindelfingen: Libertas, pp. 12–13.
41. 'Donau-Auen werden Nationalpark', *Süddeutsche Zeitung*, 28 October 1996.
42. Arbeitsgruppe der Glasmuseen und Glassammlungen unter Mitarbeit der Euregio 'Bayerischer Wald/Böhmerwald' (1994 edition) *Glas ohne Grenzen/Sklo bez hranic: Ein Führer durch die Glassammlungen und Museen zur Glasgeschichte des Bayerischen Waldes, des Böhmerwaldes und des Mühlviertels*, Deggendorf: Weiss.
43. Peter Schmitt (1997) 'Neue Chancen für "Euregio Egrensis"', *Süddeutsche Zeitung*, 27 January.
44. Jutta Seidel in conversation with the author, Edinburgh, 13 May 1996.
45. Malcolm Anderson (1996) *Frontiers*, p. 180.
46. See Michael Dauderstädt (1996) 'Ostmitteleuropas Demokratien im Spannungsfeld von Transformation und Integration', *Integration*, 4/96 (October 1996), pp. 208–23.
47. Yet exactly this, a closer co-operation between the Central and Eastern European states, and individual association with the EU on a bilateral basis, rather than membership and integration, was what Urs Leimbacher suggested. See U. Leimbacher (1991) 'Westeuropäische Integration und gesamteuropäische Kooperation', *Aus Politik und Zeitgeschichte*, B45, 1 November, pp. 3–12.

48. András Inotai (1996) 'General Attitudes towards the Europe Agreement and the European Union', in F. Mádl and P.-C. Müller-Graf (eds) *Hungary – From Europe Agreement to a Member Status in the European Union*, Baden-Baden: Nomos, pp. 13–16, p. 16.
49. Ibid.
50. Sarah Helm (1995) 'Eastern Expansion Sparks Row', *The Independent*, 16 December.
51. Alois Berger (1996) 'Europa hört nicht an der Oder auf', *die tageszeitung*, 18 December.
52. Mark Frankland (1996) 'History's Jealous Ghosts Block Doorway to East', *The Observer*, 4 February.
53. Mihály Maczonkai (1996) 'Attitudes Towards the European Union in Hungary', in F. Mádl and P.-C. Müller-Graf, op. cit., pp. 17–23, p. 18.
54. A. I. Asiwaju (1996), art. cit.
55. Michael Bosch (1996) 'Abschlußbericht', in M. Anderson and E. Bort (eds) *Boundaries and Identities: The Eastern Frontier of the European Union*, p. 106 (my translation).
56. 'Computer Worry Delays Schengen', *The European*, 28 November 1996.
57. A. Smith (1992) 'National Identities and the Idea of European Unity', p. 76.
58. Ibid.
59. Wolfgang Koydl (1996) 'Der Balkan ist Europa näher gerückt', *Süddeutsche Zeitung*, 5 November.
60. A. Bebler, *art. cit*, p. 31.
61. Ibid., p. 28.
62. See Fritz Vilmar and Wolfgang Dümcke (1996) 'Kritische Zwischenbilanz der Vereinigungspolitik', *Aus Politik und Zeitgeschichte*, 40/96, 27 September, pp. 35–45.
63. See Samuel P. Huntington (1996) *The Clash of Civilizations?*, New York: Simon and Schuster; also Carl Gustav Ströhm (1997) 'Kampf der Balkan-Kulturen', *Die Welt*, 27 January.
64. Jean-Christoph Rufin (1991) *L'Empire et les Nouveaux Barbares*, Paris: Editions Jean-Claude Lattès.
65. Perry Anderson (1996) 'The Europe to Come', *London Review of Books*, **18**(2), 25 January, pp. 3, 6–8.

Chapter 8

The Mediterranean: Europe's Rio Grande

RUSSELL KING

Introduction

Although from a European and indeed a global perspective Europe's eastern frontier has attracted the lion's share of attention over the past few years, the Mediterranean frontier remains in many respects the most problematic fringe of Europe. In fact the problematic nature of the Mediterranean frontier has, if anything, become more complex since 1990, with part of the former east–west divide running down into the Adriatic and Balkan regions, combining east–west with north–south tensions. However, unlike the old Iron Curtain, the Mediterranean is a broad, flexible maritime space across which a good deal of movement – of shipping, trade and people – has always taken place. Historians such as Braudel have shown how the Mediterranean Basin expresses a contradictory character as a region of both unity and division, of centripetal and centrifugal forces.[1] A long-standing physical and cultural unity has been repeatedly stressed by geographers,[2] whilst ethnic, linguistic and religious diversity and geopolitical fragmentation are also recurrent themes, especially in recent decades.[3] Above all, the European Union has recently become more preoccupied with the Mediterranean dimension of its external relations. Triggers for this heightened sensitivity towards Europe's southern neighbours have been the increase in regional conflicts within or overlapping into the Mediterranean Basin (Algeria, the Gulf, Cyprus, Yugoslavia, Israel) and the perceived threat of mass migration from the southern and eastern shores. The reinforcement of the EU's Mediterranean policy since 1990 reflects the serious worry that the export of terrorism and uncontrolled migration might disturb the peace and prosperity of Europe itself.

In simple terms, the Mediterranean represents a sharp, even brutal divide between 'developed' Europe and the very much less-developed realms of North Africa and the Middle East. It appears that Montanari and Cortese were the first to apply the Rio Grande concept to the Mediterranean.[4] Like its American equivalent it is a stretch of water separating two entirely different economic and social systems. Montanari and Cortese go on to show that the

American Rio Grande (including its extension across the northern Caribbean) and the Mediterranean Rio Grande (from Gibraltar to the Bosphorus)
are of comparable geographical scale and separate economic and demographic blocs of roughly corresponding weight (North America and Western
Europe on the one hand, Central America and North Africa on the other).

In the Mediterranean case, the 'development divide' has many dimensions
which generally tend to interact and reinforce each other. For instance:

- It is an *economic divide* between what used to be called the 'First' and
  'Third' Worlds, where per capita GDP levels are around ten times higher
  on the northern than the southern shores.
- It is a *demographic divide* between two entirely different population
  regimes, the high-fertility regime of the southern shore contrasting with
  sub-replacement fertility on the northern side of the Basin.
- It is a *geopolitical divide* between stable, democratic Europe and the
  generally less stable and less democratic regimes of North Africa and the
  Middle East.
- This geopolitical divide also feeds off a *cultural contrast* between Christian
  Europe and the Islamic South in which the latter is often (explicitly or
  implicitly) portrayed as an exaggeratedly non-European, underdeveloped world of 'others'.
- Finally, it is a *migration frontier* separating southern regions, where the
  'push factors' for emigration remain extremely strong, from a Europe
  which, even though the channels of immigration have been largely
  closed off since the mid-1970s, remains a desirable destination.

These five dimensions are the main themes by which we can define the
contours of the Mediterranean as a frontier region of Europe, albeit one in
which the divisive function of the sea is often overstressed when compared to
the common patterns and mutual interests across the Basin. There are also
additional, more specific arenas for potential conflict: competition over
scarce resources such as water (likely to intensify as a combined result of
population growth, expansion of tourism and intensive agriculture and
increasing drought due to global warming); the need to safeguard vital oil
resources and routes; and the commitment to control the pollution of the
sea.

We shall return to the five main themes – economics, demography,
geopolitics, culture and migration – later in this chapter. First let us explore
some aspects of the complex open/closed nature of the Mediterranean
frontier.

The Mediterranean: a broad and flexible frontier

Although this chapter – reflecting the approach of the book as a whole – mainly conceptualizes the Mediterranean as a frontier of Europe, it needs to be recognized that the Mediterranean Basin is also a regional unit which has a distinguished history and a quite specific geographical personality. In the opening sentence to their important volume on the environmental future of the Mediterranean, Grenon and Batisse assert that the Mediterranean Basin is 'an outstandingly original region'.[5] Simplifying hugely, we can say that this originality derives from the characteristic Mediterranean climate of hot dry summers and cool damp winters, the landscape of rugged hills and mountains alleviated by densely settled valleys and small coastal plains, some characteristic crops (vines, olives) and rural traditions and a long history of urban settlement dating back to classical times. Both in the classical era and in the early Middle Ages, the Mediterranean – and especially Greece, Rome, Venice, Florence and, later, Istanbul – was the setting for the accumulation of extraordinary wealth, trade and arts, at a time when northern Europe languished. Whatever role is assigned to the Mediterranean in the current European (and global) economic and geopolitical system, it should not be forgotten that the region has a self-consciously glorious past. United for the first and only time under the Romans, for whom it was simply '*mare nostrum*', it has struggled to retain its inward-looking coherence ever since. Nowadays, despite some unifying elements – such as the need for pan-Mediterranean security and the desirability of monitoring the despoliation of the shared resource of the sea – the emphasis seems to be more on the Mediterranean as a dividing-line between Europe and Africa rather than as a physical, economic and cultural expression of homogeneity.

Of course, the Romans were not averse to establishing their own bound-aries, and their concept of the *limes* is highly relevant to a debate about the frontier status of the modern Mediterranean, as Rufin has shown.[6] The *limes* which separated the Roman Empire from the lands inhabited by 'barbarians' was the line of defence erected by the rich against the danger of invasion by primitive yet warlike peoples who would destroy with their poverty the wealth and civilization of the West. It is not difficult to see where this line is being drawn at the present time: along the North African shore and around the Turkish coast.

The Roman *limes* was also a flexible concept: a closed/open strategy where there were zones of exchange and buffers ('marches') surrounding the Empire, exactly as there are today with the EU. Hence we find that the semantics of the term 'frontier' as a physical, tangible boundary or barrier have changed as functional frontiers become geographically dissociated, less visible but no less powerful. Around Europe, in the Mediterranean and elsewhere, we have multiple frontiers: fiscal frontiers, trading frontiers, investment frontiers, migration frontiers, citizenship frontiers, cultural (especially ethno-religious) frontiers, and so on. Above all we have a shift – as other chapters in this book show – from linear, territorial boundaries to ethnic and economic borders; from external to internal control; and, at the same time, from external perimeters to 'remote' control via buffer-states, tariffs and visa regimes.

Let us now turn to the various ways in which the Mediterranean can be cast as a frontier between Europe and 'beyond'.

The Mediterranean as an economic divide

This is easy to demonstrate through the availability of plentiful data on socio-economic variables such as employment, GNP and GDP per capita and human development. These data are best presented in tabular form, supported by a few remarks about the general patterns exhibited and the nature of the statistics. The countries are grouped into sets: southern EU or 'north Mediterranean' (France, Italy, Spain, Portugal, Greece); other 'north shore' countries (Malta, Cyprus, Albania); east shore countries (Turkey, Syria, Lebanon, Israel); and the south shore or North African countries (Egypt, Libya, Tunisia, Algeria and Morocco). These groupings are not entirely unproblematic. France is not normally regarded as a Mediterranean country although it has a significant stretch of Mediterranean coast. The categories violate Turkey's aspirations to be a 'north shore', European country. There are no complete data yet for the successor states to Yugoslavia. Finally, an initial glance at Table 8.1 suggests that Albania and Israel should swap places since the former is more characteristic of the southern and eastern Mediterranean whilst the latter is more 'European'.

Let us draw out a few more specific comparisons from Table 8.1, looking first at the two 'wealth' columns. Because of the way it is measured (GDP plus net factor income from abroad), GNP 'spreads' the statistical distribution of wealth between countries over a wider range than real GDP per capita, which has been standardized by 'purchasing power parities' (PPP) in order to take into account countries' varying costs of living. Hence France's GNP per capita is no less than 34 times the figure for Egypt, yet the real GDP per capita differential is only 5 times. Perhaps more typical (in that they represent comparisons between geographically close countries) are the paired cases of Spain and Morocco, and Italy and Tunisia. Spain's GNP per capita is 13 times Morocco's, its GDP ratio to Morocco 4 times. The respective figures for Italy and Tunisia are 11.5 and 5.5 times. Finally, on the wealth criterion, if we take the population-weighted average per capita GDP for the five Mediterranean EU countries ($15,565) and compare it with the weighted average

Table 8.1 Socio-economic indicators for Mediterranean countries

	GNP per capita (US$) 1993	Real GDP per capita (PPP) 1993	HDI 1993	% of labour force in (1992)		
				agriculture	industry	services
Mediterranean EU						
France	22,490	19,140	0.94	6	29	65
Italy	19,840	18,160	0.91	9	32	59
Spain	13,590	13,660	0.93	11	33	56
Portugal	9,130	10,720	0.88	17	34	49
Greece	7,390	8,950	0.96	23	27	50
Other Europe						
Malta	7,970	11,570	0.89	3	28	69
Cyprus	10,380	14,060	0.91	15	21	64
Albania		2,200	0.63	56	19	25
Eastern Shore						
Turkey	2,970	4,210	0.71	47	20	33
Syria		4,200	0.69	23	29	48
Lebanon		2,500	0.66	14	27	59
Israel	13,920	15,130	0.91	4	22	74
Southern Shore						
Egypt	660	3,800	0.61	42	21	37
Libya		6,125	0.79	20	30	50
Tunisia	1,720	4,950	0.73	26	34	40
Algeria	1,780	5,570	0.75	18	33	49
Morocco	1,040	3,270	0.53	46	25	29

Source UNDP (1996)

for five south and east Mediterranean countries ($4,150 for Turkey, Egypt, Tunisia, Algeria and Morocco), the ratio becomes 3.75. This comparison has particular realism because both five-country blocs have an almost identical population (around 180 million) and the two groups represent the main origins and destinations for recent trans-Mediterranean migration.[7]

As students of development are at pains to point out, wealth is often a poor measure of development, if by 'development' is meant a human as opposed to a purely economic process. In its annual *Human Development Report*, published since 1990, the UNDP has refined its measurements of human development and demonstrated the fallacies of 'the mismeasure of human development by economic growth alone'.[8] Hence true development is much less about a country's total accounted wealth (itself a tricky concept to grasp and to measure), and much more about distribution of income, access to jobs, quality of life, health, education, gender equality, human rights and good governance.[9] In Table 8.1 the human development index (HDI) is a composite measure made up in equal weight of three variables: real per capita GDP, life expectancy and education (measured by a combination of adult literacy and the mean number of years of schooling). The figures show the divide in human development is less sharp than for GNP or GDP. The data also show a subtle reordering of certain countries; for example, Spain has a higher HDI, but lower GNP/GDP per capita, than Italy. Although the relative gap between the northern shore EU countries and the rest is narrower on the HDI criterion, the division is nevertheless still significant between two standards of human development – Europe plus Israel and minus Albania; and the rest of the Mediterranean Basin.

The final section of Table 8.1 sets out the distribution of the labour force between the three major employment sectors. These data show, again with some exceptions, a contrast between a 'European' Mediterranean which has progressed to an economy dominated by service employment (generally over 50 per cent) and, to a lesser extent, by industry (generally around 30 per cent); and, on the other hand, the southern and eastern Mediterranean countries (including Albania) where agricultural populations remain important (notably Albania 56 per cent, Turkey 47 per cent, Morocco 46 per cent, Egypt 42 per cent). However, in Syria, Lebanon, Libya, Algeria and Tunisia, farm workers are exceeded by both the industrial and service sector labour forces. Interestingly the two countries with the lowest agricultural and the highest service populations are Malta and Israel, but both can be regarded as special cases.

Behind the employment figures lie broader social processes like modernization and urbanization. There is no space to explore these processes here but it is important to have some idea of the pace and scale of socio-economic change in different parts of the Basin. Table 8.2 demonstrates that, by and large, the non-European Mediterranean has been experiencing fast structural change since the 1960s, with regard to both improvements in HDI values and the restructuring of employment away from agriculture. Changes have also been marked in southern Europe but, measured against higher base levels, they tend to appear less dramatic. Historical comparisons are often dangerous, but the evidence suggests that countries like the Maghreb states are now roughly at the stage that southern European countries such as Spain, Portugal and Greece were in the late 1950s or early 1960s. To pinpoint this comparison, the 1992–3 data for Morocco are virtually identical to those for Greece in the early 1960s (Table 8.2).

The Mediterranean as a demographic frontier

If the economic division across the Mediterranean is very clear, the demographic gradient is, if anything, even sharper: a demographic fault-line between the static, even shrinking populations of Western Europe and the high-fertility, 'Third World' demographic regimes of North Africa and the Near East. Table 8.3 presents some relevant indicators.

Table 8.2 Human Development Index and employment structure in Mediterranean countries: early 1960s and early 1990s

	HDI		Agriculture (%)		Industry (%)		Services (%)	
	1960	1993	1965	1992	1965	1992	1965	1992
Mediterranean EU								
France	0.85	0.94	18	6	39	29	43	65
Italy	0.76	0.91	25	9	41	32	34	59
Spain	0.64	0.93	34	11	34	33	32	56
Portugal	0.46	0.88	38	17	30	34	32	49
Greece	0.57	0.90	47	23	24	27	29	50
Other Europe								
Malta	0.52	0.89	8	3	41	28	51	69
Cyprus	0.58	0.91	40	15	27	21	33	64
Albania		0.63	69	56	19	19	12	25
Eastern Shore								
Turkey	0.33	0.71	75	47	11	20	14	33
Syria	0.32	0.69	52	23	20	29	28	48
Lebanon		0.66	29	14	24	27	47	59
Israel	0.72	0.91	12	4	35	22	53	74
Southern Shore								
Egypt	0.21	0.61	55	42	15	21	30	37
Libya		0.79	41	20	21	30	38	50
Tunisia	0.26	0.73	49	26	21	34	30	40
Algeria	0.26	0.75	57	18	17	33	26	49
Morocco	0.20	0.53	61	46	15	25	24	29

Source UNDP (1996)

The data show that population growth is slowing everywhere except in Malta, Cyprus, Syria and Lebanon, all relatively small countries. In the Mediterranean EU countries it is slowing in the mid and late 1990s to zero growth, especially in Italy, Spain and Portugal. In the eastern and southern Mediterranean, current annual population growth rates are between 1.8 and 3.4 per cent. Turkey, Tunisia and Morocco are leading the fast decline in the rate of growth in the 'underdeveloped' part of the Basin. Israel's continuing buoyant demographic growth is fed by high levels of immigration, especially from the former Soviet Union, as well as a relatively high fertility rate.

Similar contrasts are evident when we examine the total fertility rate (TFR) column in Table 8.3. This refers to the average number of children per woman over the reproductive life cycle: the figure must be above 2 to ensure the long-term replacement of the population. The countries neatly divide themselves into four groups based partly, but not entirely, on the groupings set out in the table. The first group, the southern EU countries, are countries of extremely low fertility, well below replacement level. If these hyper-low fertility levels persist Italy, followed quickly by Spain, Greece and Portugal,

Table 8.3 Selected demographic data for Mediterranean countries

	Population (m)	Average annual growth (%)		TFR	Ageing index		
	1993	1960–93	1993–2000	1992	1985	2000	2020
Mediterranean EU							
France	57.5	0.7	0.4	1.7	61.1	82.6	134.4
Italy	57.1	0.4	0.0	1.3	65.0	114.9	168.2
Spain	39.5	0.8	0.1	1.2	53.1	89.6	119.6
Portugal	9.8	0.3	0.0	1.6	50.9	79.6	111.4
Greece	10.4	0.7	0.3	1.4	62.4	102.1	131.5
Other Europe							
Malta	0.4	0.4	0.6	2.1	41.0	65.7	117.2
Cyprus	0.7	0.7	1.0	2.5	42.3	50.6	83.6
Albania	3.4	2.3	1.0	2.9	15.3	23.0	43.1
Eastern Shore							
Turkey	59.6	2.4	1.8	3.4	11.7	20.8	43.8
Syria	13.7	3.4	3.4	5.9	5.7	6.0	7.1
Lebanon	2.8	1.3	2.3	3.1	13.6	19.6	29.7
Israel	5.3	2.8	2.1	2.9	27.0	33.7	50.4
Southern Shore							
Egypt	60.3	2.4	2.0	3.9	9.6	13.0	22.5
Libya	5.0	4.1	3.4	6.4	5.0	6.4	8.7
Tunisia	8.6	2.2	1.8	3.2	9.7	13.7	21.0
Algeria	26.7	2.8	2.2	3.9	8.0	8.5	12.4
Morocco	25.9	2.5	1.9	3.8	9.2	10.7	16.9

Source Livi-Bacci and Martuzzi Veronesi (1990); UNDP (1996)

will start to experience quite sharply declining absolute populations in the early decades of the twenty-first century. Population analysts now speak of a specifically southern European demographic regime, for Italy (in the late 1980s) and Spain (in the 1990s) have the lowest fertility levels ever recorded anywhere in the world.[10] There is no space here to explore in detail the reasons for this new, 'post-demographic transition' southern European pattern. Very briefly, some of the probable factors are: the high levels of university education, but the difficulty of starting a career in a tight and hierarchically structured labour market (hence the postponement of marriage and family formation); the trend for young women to increasingly opt for a career rather than a family; the lack of public childcare support making the combination of a career and child-rearing very difficult; and the still-sharp gender stereotypes preventing men from lending support in childcare or home-making.[11]

The second group of countries – Malta, Cyprus, Albania, Israel – have TFRs of 2.1–2.9. These are small countries of medium fertility but otherwise have little in common. Malta approximates most closely to the new southern European model discussed above. The situation in Cyprus is complicated by ethnic partition and the recent settlement of high-fertility Turks from mainland Turkey in the northern part of the island. Israel's fertility is boosted by the immigration of many young families and the pro-natalism of Orthodox Jews, whilst Albania's fertility appears to be falling rapidly.

The third group of countries are medium-high fertility countries with TFRs of 3.1–3.9. This is a large group of southern and eastern Mediterranean states with a combined population of 184 million in 1993. Fertility is falling quite fast[12] but there is still considerable growth momentum: the populations remain weighted towards the younger cohorts, with around 40 per cent under 15 years of age. The population of these countries in the 1990s is still growing at around 2 per cent per year. This is the group of countries with the highest migration potential over the next few decades.

Finally there are two countries – Syria and Libya – with very high TFRs, where the average woman has around six children and the population grows at 3.4 per cent per year. This means that the population of these countries will double in just twenty years.

In many respects the most remarkable statistics on contrasting population structures within the Mediterranean Basin are those on age distribution. The final three columns in Table 8.3 show how the 'ageing index' (the ratio of those over 65 to those under 15) will develop over the period 1985–2020. The data are based on UN Population Division medium-variant scenarios and obviously rely on potentially errant assumptions regarding fertility behaviour in the years between now and 2020; the elderly population is easier to predict since they are already born. In 1985, all southern European EU countries had elderly populations (65+) which were between half- and two-thirds the size of their under-15 cohorts. During the 1990s, first Italy and then Greece saw more over-65s than under-15s in their populations, and this trend is projected to increase strongly by 2020. Particularly dramatic will be the ageing of the Italian population: in 2020 there will be 12.2 million over-65s compared to 7.25 million under-15s. This represents an almost exact reversal of the situation in 1985 when there were 11.2 million under-15s and 7.28 million over-65s.[13] By contrast, the ageing of the population structures of the southern and eastern Mediterranean countries is proceeding only slowly. Even Tunisia, which is in the vanguard of North African fertility decline, will have an index of only 21 per cent in 2020 (of course these figures could increase if either life expectancy increased faster than expected and/or fertility fell more sharply than predicted).

The combination of the economic and the demographic divide running through the Mediterranean forms the Rio Grande effect which is encapsulated in the title of this chapter. Economic and demographic variables create the setting for mass trans-Mediterranean migration which appears to be Europe's greatest fear concerning its southern margins and which we will

examine towards the end of this chapter. Uneven development, the demographic contrast between 'First World' and 'Third World' population
regimes, and migratory pressure are also intimately linked to the geopolitical
sensitivity of the Mediterranean Basin. But they are by no means the only
influences on Mediterranean geopolitics, as the next section will show.

Mediterranean geopolitics

The geopolitical significance of the Mediterranean derives from some simple
facts of history and geography. The region lies at the junction of three
continents: Europe, Africa and Asia. Twenty nation-states, a dozen languages, three major religions and a diversity of cultures and political systems
abut its shores. Add to this the powerful gradients of uneven development
and differential population growth discussed above, plus the strategic location of the Basin as a funnel for major oil routes, and the potential for
geopolitical conflict is clear.

Two other key points need to be made by way of introduction to the
geopolitical significance of the Mediterranean. First, removal of the east–
west ideological and political divide has refocused and intensified attention
on the Mediterranean as an axis of tension. Second, the removal of the US–
Soviet superpower balance in the Mediterranean, during which naval and
other forces effectively neutralized each other, stifling most conflicts, has
allowed tensions to erupt which previously lay dormant under the 'dead
hand' of the Cold War. Most of these conflicts are regional (Israel/Palestine/
Lebanon, the Balkans) or local (Cyprus, Algeria) rather than global, but the
potential for them to spill over into neighbouring countries, or to 'spill in'
from outside (as with the Gulf War), makes the disturbances of more than
restricted interest.

Let us now examine Europe's developing institutional and geopolitical
relationship with the Mediterranean – a relationship which can be described
as moving progressively from pragmatism to partnership,[14] and which has
developed against the background of mounting fears regarding security and
mass immigration. The starting point for this examination is the increasing
Mediterranean character of the EU itself, particularly as a result of its second
enlargement to include Greece, Portugal and Spain in the early–mid-1980s.
Prior to this, and as early as 1964, the Italian government had issued
statements pushing for an overall Common Market policy towards the
Mediterranean region. By 1972, the Community had put in place its so-called
'Global Mediterranean Policy' (GMP) which was designed to secure some
political stability in the region on the basis of free trade, EC financial aid for
economic development and associate member status for a number of Mediterranean countries, starting with Greece and Turkey in the early 1960s.

The geopolitical setting of the GMP contained a number of important elements. First, it was to be seen against a history of trans-Mediterranean colonial links, above all between France and the Maghreb states and between Italy and Libya. Second, it developed against a background of specific crises, notably the Arab-Israeli wars of 1967 and 1973. Third, it represented an ambitious strategy of providing an alternative to the bipolar superpower influence in the region. Economically, it was a progression from the fact that the Mediterranean had become an important trading bloc for the Community.

But there were also important constraints on its success.[15] The first Community enlargement, in 1973, brought British interests into play. Whilst the UK had its own Mediterranean colonial legacies in Gibraltar, Malta and Cyprus, it also had other priorities, notably its concern that preferential trade agreements with Mediterranean states might compromise UK–US relations through infringements of GATT, and its wider overseas interests including its links to former colonies. Second, Yom Kippur made the Community appreciate the difficulty of a uniform or even-handed policy towards the various Mediterranean and Middle Eastern states.[16] Third, there was the rapidly changing context of southern Europe itself: the legacy of long-term fascist rule in Spain and Portugal (and Greece during 1967–74); the restoration of full democracy in these countries in the mid-1970s; the long-running Cyprus problem, made acute by the Turkish action in 1974 to divide the island.

In 1990, the Community agreed a new Mediterranean policy which reflected the full consequences of the southern enlargement as well as other major European processes such as the evolving progress towards a Single Market and the democratization of Eastern Europe. This new policy was both more broadly based and more pro-active than the GMP. Above all it reflected a growing fear in the corridors of European power that

> if Europe did not concern itself directly with the problems of its southern neighbours, then these problems would be exported northwards in the form of population movements and serious threats to stability and security.[17]

Amongst key measures of the new policy were support for structural adjustment in the non-European Mediterranean states, direct investment by both European private companies and Community financial assistance, preserving and where possible enhancing access for Mediterranean products to Community markets, and strengthening the amount of political and economic dialogue across the Mediterranean. The Community allocated 4.4 billion ECU in grants and loans for the period 1991–6 to support Mediterranean development. This sum was targeted both at individual countries (Morocco, Algeria, Tunisia, Egypt, Israel, Jordan, Lebanon, Syria) and at broadly based assistance and networks such as the Med-Urbs, Med-Invest, Med-Campus and Med-Media initiatives designed to increase economic, technical and cultural co-operation between countries in the Community and those in the Mediterranean region. By 1994, however, it was acknowledged that these new policies and instruments had not been very effective and at the end of that year the creation of a 'Euro-Mediterranean Economic Area' was announced as a step towards realizing the ambitious objective of an EU–Mediterranean free trade area by the year 2010.

This brings us to the latest stage in Europe's developing relationship with the Mediterranean – the establishment of the 'Euro-Med' partnership at the Barcelona conference of November 1995, attended by leaders from the 15 EU states and 12 Mediterranean countries (Albania, Libya and the Yugoslavian successor states were the absentees). Behind the rhetoric and the elaborate language of compromise of the declarations of this meeting – about economic co-operation, cultural exchange and respect for human rights and fundamental freedoms – there lay a more simple, brutal message: trade and aid, but not migration. According to an article in *The Economist*, the meeting was united by fear:

> in Europe the fear of Islamic fundamentalism on the southern shore and of immigrants fleeing civil strife or poverty; in North Africa and the Middle East the fear that Europe will turn inwards and expel migrants, who will return to make matters worse at home.[18]

Throughout the Barcelona debate, migration was linked with security: innocent people (whose only 'crime' is their poverty and keenness to migrate to improve their lot) were bracketed with terrorism and the smuggling of arms and drugs. The logic of the Barcelona meeting was to 'buy security' through economic development and to create conditions whereby emigration from the southern and eastern shores will be stemmed. In addition to reaffirming the objective of a Euro-Mediterranean free trade zone by 2010, the main financial commitment is 4.7 billion ECU of aid offered to the 12 non-EU Mediterranean states during 1996–9, boosted by loans from the European Investment Bank and by bilateral aid from the EU's Member States. This clearly represents an enhancement of the commitment agreed under the 'new Mediterranean policy' of the early 1990s noted above, and brings the total financial package for the period 1991–9 to 9.1 billion ECU.

This may sound impressive, but there are other ways of evaluating these figures. An analysis by *The Economist* showed that in 1994 EU aid to the Mediterranean countries was half the level allocated to Central and Eastern Europe.[19] Whilst it is true that this differential becomes two-thirds after the Barcelona conference ($6 billion to the Mediterranean, $9 billion to Eastern Europe during 1996–9), these fractions must be recalculated according to

Table 8.4　The Mediterranean and East European countries: a comparison of EU support

	Mediterranean countries[1]	Eastern Europe[2]	EU 12
Population, 1993, millions	204	109	348
Estimated population, 2010	297	116	376
GDP, 1993 $ billions	362	224	6,785
GDP per head, 1993, $	1,746	2,057	19,485
EU budget commitment, 1994, $ millions	563	1,172	
EU budget commitment per head, 1994, $	2.76	10.75	
Estimated EU budget commitment per head per year, 1996–9, $[3]	6.82	20.27	

Notes:　[1] Algeria, Cyprus, Egypt, Israel, Jordan, Lebanon, Malta, Morocco, Syria, Tunisia, Turkey
　　[2] Albania, Bulgaria, Czech Republic, Estonia, Hungary, Latvia, Lithuania, Poland, Romania, Slovakia, Slovenia
　　[3] Based on $6 billion allocation to the Mediterranean and $9 billion to Eastern Europe for 1996–9, and on estimated population totals for 1997
Source Partly after *The Economist*, 2 September 1995

the populations of the two regions. The Mediterranean grouping contains nearly twice the population of Eastern Europe and there are wide differences in projected demographic growth rates. Hence the real differential of EU support is three to four times higher in Eastern Europe than the Mediterranean. Table 8.4 sets out these comparisons in more detail.

There are also doubts about the true degree of economic co-operation and reform that can be achieved. The problem here is an asymmetrical balance of trade and of trade dependency. The Mediterranean countries are very reliant on the EU as an export market (more so in fact than are the countries of Eastern Europe), yet they have a trade deficit with the EU (of around 10–12 billion ECU in 1993 and 1994) which is not particularly reliant on the Mediterranean as a major export outlet. Trade orientation to the EU is particularly high in the Maghreb (Tunisia, Morocco and Algeria sent respectively 80, 69 and 68 per cent of their exports to the EU in 1994); it lessens towards the eastern Mediterranean where Middle Eastern and other markets become more significant. The trading problem is compounded by the fact that the Mediterranean countries export mainly agricultural products, textiles and other low-technology goods. These are often in surplus and/or compete with the output of the southern EU states. The 'free trade by 2010' agreement is therefore mainly based on eliminating trade barriers for manufactures; trade in agriculture will be liberalized 'only as far as the various agricultural policies will allow' – in other words the Common Agricultural

Policy and the protection of Europe's farmers will remain the dominant
policies (despite continuing CAP reforms), much to the frustration of countries like Egypt and Morocco.[20]

Equal scepticism can be applied to an evaluation of the interlinked political
issues of democratic reform, human rights, territorial integrity and arms
limitation. The Barcelona declaration paid lip-service to human rights but
did not make compliance a condition for receiving aid or loans. This
recognizes the *realpolitik* of dealing with regimes whose repressive policies
and tolerance of (or failure to eradicate) terrorism and extreme dogma are
not going to change overnight. It also recognizes the conflicting positions
and interpretations of issues of arms proliferation and territorial self-
determination held by Israel, Syria and the Palestinian delegation.

Finally, there are some wider political contexts of the Mediterranean to be
mentioned. The security challenges of the Mediterranean region extend well
beyond Europe to the global level, as they have done for many decades, if not
longer. However, the new security architecture of the post-Cold War era
introduced new uncertainties in the void left by the superpowers, as was
noted earlier. The attempt to draw up a Conference on Security and Co-
operation in the Mediterranean (CSCM) in 1990, launched by Spain and
Italy, has been an initiative whose long-term effectiveness has been variously
interpreted: Anderson and Fenech see it as a 'potentially unifying forum for
the region as a whole'[21] with 'the capacity to build bridges of an economic,
political and strategic nature between the north and south of the Mediterranean',[22] whilst Joffé regards it as too complex and cumbersome to be of
much use in practice – for him more specific spheres of co-operation, such as
between the EU and the Maghreb, offer better prospects for tackling regional
problems.[23] Certainly, the idea of the CSCM has yet to be vigorously pursued
at a political or diplomatic level; hence it remains Mediterranean multi-
lateralism on paper rather than in reality.

On the other hand, the fact there is only one global superpower means that
the geopolitical views of the Mediterranean held by the United States must be
part of any analysis of the region. The US global view contains many angles
on the Mediterranean: southern Europe and Turkey as the southern flank of
NATO (yet to redefine its role following the demise of the Warsaw Pact); the
Mediterranean as a corridor of access to both the oilfields and trouble-spots of
the Middle East and the Gulf; and the existence of the Mediterranean as a
global North–South fracture zone, demarcated by a boundary running east–
west through the southern waters of the basin, south of the Balearics,
Sardinia, Sicily, Malta and Crete. Holmes asserts that (southern) Europe and
the United States have strong reasons for co-operation in the Mediterranean:
the US needs southern European co-operation to guarantee access to the
Middle East; southern Europe needs the US simply because it is the only power

capable of effective involvement in the area.[24] Hence America has, to use Lesser's words, 'rediscovered the Mediterranean' in the new global geopolitical map.[25] The basin is seen as an extension of the European environment and transatlantic relations, as an aspect of Middle Eastern geopolitics (notably the Arab-Israeli peace process), and as a potential source of local conflicts. In addition, migration, economic development and their implications for regional security and prosperity are also at the heart of the emerging 'American debate' about the Mediterranean.[26] Both Anderson and Fenech and Lesser describe the Mediterranean as the entrance to a new 'arc of crisis' which starts with the potential flashpoint of Algeria, continues past the *bête-noir* of Libya and then passes through the Middle East and Gulf regions to circle round via South-Central Asia to the Balkans and East-Central Europe.[27] Nevertheless it must also be appreciated that many of the tensions of the Mediterranean area are being played out largely *within* states such as Algeria, Egypt, Lebanon, Israel/Palestine, Cyprus and Bosnia; and the objective of both American and European policy has been to keep these conflicts contained at the local scale.[28]

The Mediterranean as a cultural divide

In Chapter 1 of this book, Malcolm Anderson outlined a variety of approaches to the study of frontiers which range from the relatively simple view that they are epiphenomena whose role and location depend on the core characteristics of the state they encircle to more complex notions that frontiers are markers of identity. In this latter interpretation, frontiers are part of beliefs (and myths) about national unity – or, in the case of the EU, about a kind of supranational European unity. They can be used not only to define what, and who, lie within the chosen territory but also, more prejudicially, the kinds of peoples who are to be excluded. Religion and culture have become 'belief frontiers' for Europe, demonstrating that 'holy selfishness is the supreme law of any community' and clearly positioning the Islamic south and east Mediterranean as beyond the European *limes*.[29]

According to Rufin, the southern Mediterranean (and beyond) is becoming more and more what it was in the early nineteenth century: a *terra incognita* where the European 'white man' can no longer go.[30] The ethno-religious criteria are all too clear: Europe is white, 'civilized', democratic, Christian (Protestant, Catholic, Orthodox in the case of Greece); not Arab, Ottoman, Islamic or Slav. From the perspective of the 'new Europe' and especially from the perspective of the southern EU states, this cultural imperialism is based less on traditional markers such as skin colour or colonial, 'Third World' status, and much more on a new trans-Mediterranean racism directed against a Muslim world which is poorly

understood despite its geographical closeness, and in which fundamentalism, abuse of human rights, suppression of women and explosive
population growth can all be easily demonstrated. This allows the Muslim
Mediterranean to be rejected wholesale as 'barbarian'.[31]

It is not difficult to explode this Eurocentric myth, and Edward Said's
Orientalism (1978) remains a fundamental starting point for such a decon-
struction. Key points in this important debate, insofar as it affects the falsely
imposed cultural apartheid of the Mediterranean, include the following.
First, there is the arbitrary annexation by Europe of Christianity, despite its
Middle Eastern origin. Second, there is a parallel annexation of Hellenic
culture by Europe. Hence the history of 'Western thought' is conventionally
traced back to Ancient Greece and 'came of age' with its reappropriation by
the Renaissance.[32] Third, the present north-west/south-east cultural division of the Mediterranean is projected backwards to recall powerful historic
events and myths from the past: the Crusades, the arrival of the Ottomans at
the gates of Vienna in 1683, the Spanish *reconquista*, etc. The European-
ization of the southern Mediterranean – colonial rule throughout North
Africa and the curious fact that Algerians were constitutionally French
citizens before 1962 – is conveniently forgotten.

Fourth, the cultural divide hypothesis rests on a false homogenization of
both European and Islamic–Oriental cultures. Neither have much cultural,
religious or linguistic unity when examined closely. Such mythical identities
rest on notions of the 'purification of (European) space' and a 'geography of
rejection' which shuts out those who threaten the purity.[33] Whilst the
processes of rejection are very real – as will be seen in the next section on
migration – the notion of purity as applied to territories and cultures is
largely false. Places and cultures are the product of a long history of
connections to other places and cultures. In any case the very notion of
culture is a contested concept. Stuart Hall sees cultures not as settled,
enclosed and internally coherent but as meeting-places, contact-zones based
on the intersection of different traditions, constantly changing and with no
simple, single origin.[34] Migrations within and into Europe since the 1950s
are thus merely the latest stages in a long process of European hybrid-
ization.

The inescapable conclusion to this fourth key point is that the
dichotomization of the Mediterranean into a Christian, 'civilized', 'developed'
north-west and a Muslim, 'underdeveloped', 'barbarian' south-east is a
pernicious oversimplification. Quite apart from the confused situation in the
Balkans, there are significant Christian communities in Lebanon, Israel/
Palestine and Egypt, whilst the Muslim presence in the EU now amounts to
an estimated 7 million, including 2 million in France (mainly North Afri-
cans), 2 million in Germany (mainly Turks), 1 million in the United Kingdom

(mainly Pakistanis and Bangladeshis), 750,000 in the Benelux countries (mainly Turks and Moroccans) and perhaps 700,000 in southern Europe (Moroccans, Tunisians, Albanians, Senegalese, etc.).

The Mediterranean as a migration frontier

To a certain extent, all of the previous Mediterranean frontiers described in this chapter – economic, demographic, geopolitical, cultural – coalesce into one reality and one fear: the Mediterranean as a setting for past, present and future northward migrations into Europe. The 'new migrations' across the Mediterranean are the new version of Orientalism: building a Fortress Europe and a new sense of (West) European identity by creating a new set of 'others'.[35] Migration redistributes (and creates) cultures, and the EU appears to be deeply afraid of this.

Thirty or forty years ago the situation was very different. During the 1950s and 1960s much of the Mediterranean Basin (to be precise, Spain, Portugal, Italy, Greece, Yugoslavia, Malta, Cyprus, Turkey, Tunisia, Algeria and Morocco) shared a uniform function as a migration 'reservoir' for the industrial countries of north-west Europe. As Figure 8.1 shows, the 'migration frontier' (immigration countries to the north, countries of emigration to the south) ran through southern Europe, not through the Mediterranean. The critical mechanism driving this mass northward migration of Mediterranean labour was the need for relatively cheap and unskilled workers to ease the tight labour markets in post-war Europe's booming industrial economies. Any worries about cultural differences between host and sending countries were sidelined by the argument that most of the migrant workers would return home in a few years' time. Of course, most did not return; they stayed on, coping with racism and prejudice, and helped to contribute a distinct Mediterranean character to European cities and populations.

By the 1980s, Europe's migration frontier had shifted southwards to run along the coast of North Africa and the eastern Mediterranean, including Albania as a significant source country after 1990 (Figure 8.2). The Mediterranean can now be considered Europe's Rio Grande, but is much more difficult to control and police than the US–Mexican border. The impossibility of completely sealing the 'soft underbelly' of the EU to 'unwanted' immigration reflects both objective geographical conditions – the impracticality of patrolling so much open sea and empty coastline – and the irresistibly strong pressures driving cross-Mediterranean migration. These pressures exert particular force where the distance between potential sending and receiving countries is very short. Across these 'mini-Rio Grandes' migrants have been known to swim from Morocco to Spain, or from Albania to Corfu, repeating the experience of the illegal 'wetbacks' who swam the Rio Grande to enter

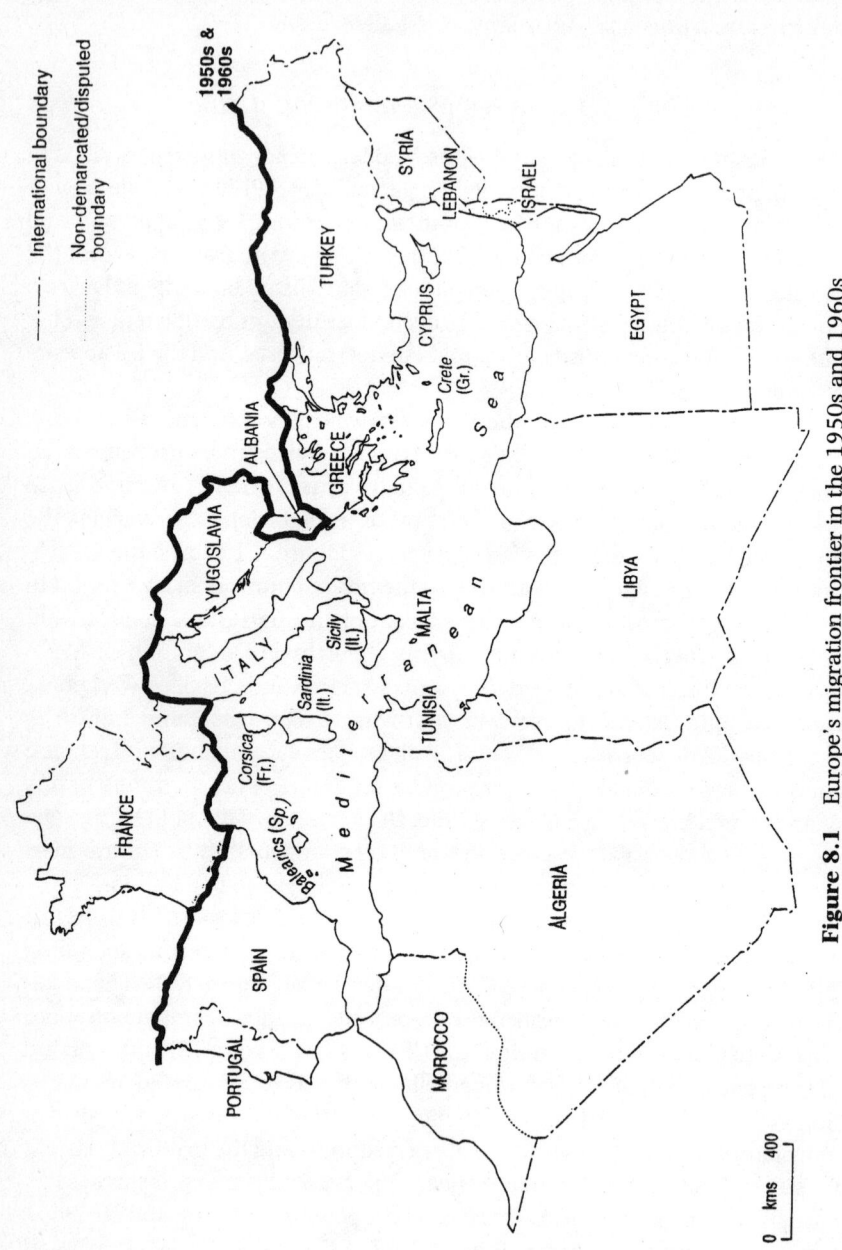

Figure 8.1 Europe's migration frontier in the 1950s and 1960s

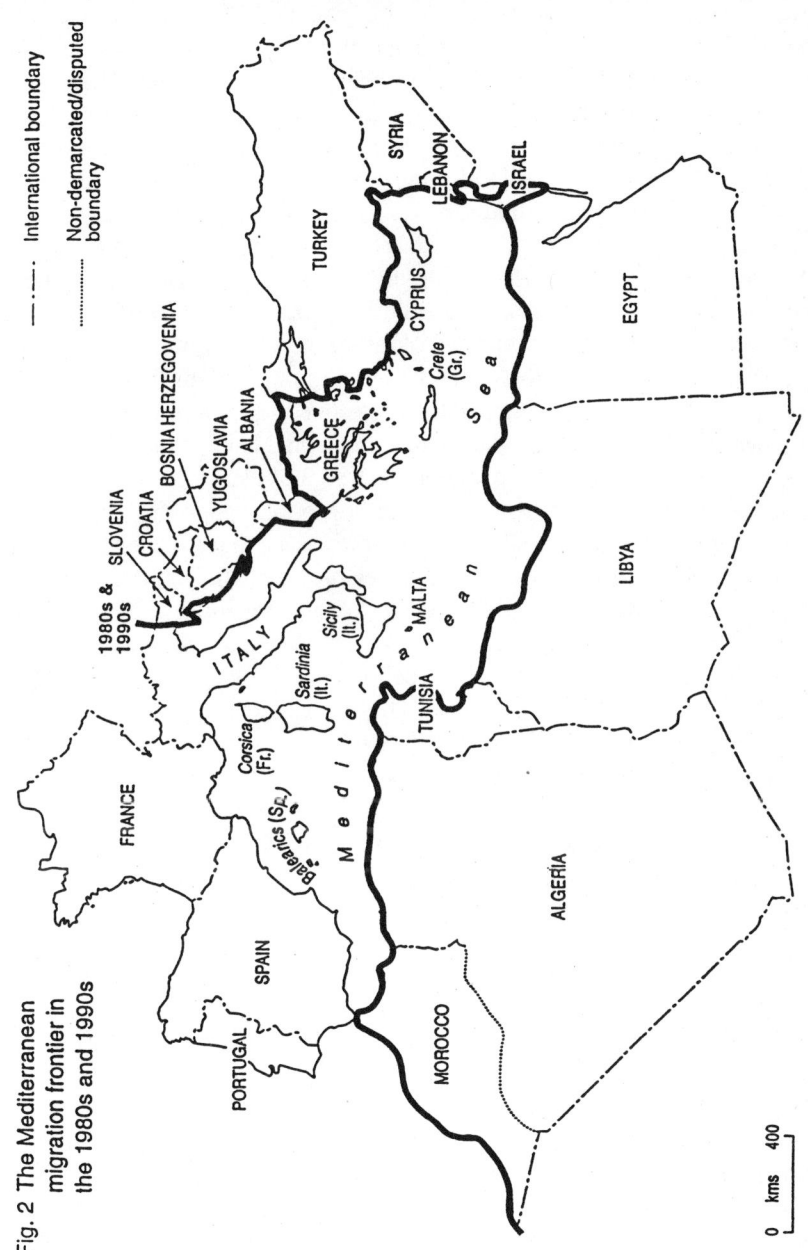

Fig. 2 The Mediterranean migration frontier in the 1980s and 1990s

International boundary

Non-demarcated/disputed boundary

Figure 8.2 The Mediterranean migration frontier in the 1980s and 1990s

the United States. Elsewhere in the Mediterranean the smuggling of migrants is active between Tunisia and Sicily, between Albania and the heel of Italy and between Turkey and nearby Greek islands. The mountainous land frontier between Albania and Greece has also proved impossible to control, with the result that an estimated 300,000 Albanians – one-tenth of the country's population – migrated to Greece in the early 1990s.[36]

It is common to explain the trans-Mediterranean migrations of the 1980s and 1990s largely in terms of a cocktail of 'push factors' emanating from the countries of origin: population growth, unemployment, low incomes, poverty, environmental degradation, war, ethnic strife. Certainly, these factors provide the structural context for migration into Europe, especially southern Europe, from the other side of the Mediterranean. Since the 1970s, powerful migratory currents have connected Morocco with Spain and Italy, Tunisia with Sicily and Egypt with Greece. Over the longer term, it is difficult to ignore the relevance of a fact like the projected 133 million increase in the number of working-age people in the southern Mediterranean countries (Turkey clockwise to Morocco) between 1985 and 2020.[37]

But the real explanations and mechanisms of migration are much more complex and wide-ranging. It is important to realize that the Mediterranean is the friction plane between two wider regions: the whole of Europe to the north, and the whole of Africa and much of Asia to the south and south-east. The geography of air routes and previous colonial and religious links also connects Latin America strongly to southern Europe. Additional relevant factors include the historical context of intra-Mediterranean migrations and the existence of well-established linkages between origins and destinations, the role of mass media (for example, Italian television programmes received in Tunisia and Albania), ease of entry to southern as opposed to northern EU countries, the emergence of Mafia-like organizations to smuggle people to remote coasts and islands and the complex relationship between immigrants and the southern European economy and labour market.[38]

The last of these factors deserves special attention. Despite a common perception that the recent cross-Mediterranean migrations have been supply–push rather than demand–pull movements, and despite the high levels of unemployment recorded in southern Europe in recent years (notably in Spain and southern Italy), the character of the labour market in Mediterranean Europe does favour the use (and exploitation) of immigrant labour. Two particular features stand out: the seasonality of much southern European economic activity, especially tourism and specialized agriculture, requiring flexible supplies of casual labour which are unavailable locally (partly due to the low prestige of manual labour in the modernized societies of southern Europe); and the strength and dynamism of the informal sector or black economy, which feeds off immigrant labour. North African (and other) immigrants have thus come to underpin many sectors of the southern European economy – agriculture, fishing, tourism, the construction industry, personal services (domestic help, hospital and care workers) – as well as create livelihoods for themselves around the fringes of the economy as street-hawkers, market-traders, windscreen washers, etc.

The fact that so much immigration into southern Europe has been allowed (3 million immigrants are living in Spain, Portugal, Italy and Greece according to King and Konjhodzic)[39] is an indication not only of the difficulty of 'closing the fortress' against 'invaders' from the south, but also of immigrants' potential usefulness in keeping the labour market responsive to employers' needs and in holding down the overall costs of production. This also explains, perhaps, the tokenism of recent policy initiatives such as periodic regularizations in Spain and Italy since the mid-1980s and the far-from-effective repatriations of Albanians by the Greek police since 1991 (many Albanians simply re-enter by walking back over remote mountain paths).

Hence against the negative, keep-them-out, militaristic discourse of the Mediterranean as the last line of defence against the hordes of would-be migrants massing on the southern shores, there is an alternative view.[40] This view stresses that immigrants are generally an economic asset (as they have been in virtually every country in the world where labour migration has taken place), and points to the trans-Mediterranean complementarity of demographic regimes. The links between migration on the one hand and terrorism, criminality and drug-smuggling on the other hand are simply not properly proven. Of course, it cannot be denied that some immigrants are involved in these activities, but so are EU nationals, and the criminalization of the act of migration by the European definitions of illegality of movement into the EU makes migrants more vulnerable to repressive policing and guilt-by-association. And far from being the bearers of fundamentalist Islam, most cross-Mediterranean migrants from Islamic countries are broadly secular, even Western, in their ideological outlooks.

Conclusion

What of the future of Europe's Mediterranean frontier? The Mediterranean nightmare is obvious:

> North Africa disintegrates into civil wars, terrorism and religious zealotry infect Europe's cities, and Arab migrants, by the hundreds of thousands, cross the Mediterranean for a better life in an unwelcoming, racist-inclined Europe.[41]

As noted above, the EU is united by this 'fear of the South', which is also –

and hence its exaggeration – a fear of the unknown and of the misunderstood, a rejection of other cultures which are demonized by reference to distorted histories. As Didier Bigo's chapter in this book shows, politics strengthens the rhetoric, securitizing the anti-immigration, anti-South discourse at the symbolic level, whilst appreciating that in reality the southern borders can never be sealed completely.

It is clear that the EU's Mediterranean frontier is qualitatively different from the eastern frontier in a number of respects. As an open sea frontier it is more difficult to control, and beyond it lie a set of countries with economic, demographic and cultural structures that are very different from those of Eastern Europe. Elaborating on this theme, Lipietz recognizes four 'circles' of countries beyond the external frontier of the EU.[42] First, there is the old EFTA circle: a set of countries with levels of development close to or above the EU norm but also a commitment to neutrality. The removal of the East–West military dimension allowed Austria, Sweden and Finland to join the EU in 1995, leaving a residual group of just two, Norway and Switzerland. Second, there is Eastern Europe, which we can divide into two subgroups of countries: those whose reforming economies make them possible candidates for EU membership in the not-far-distant future (the Czech and Slovak republics, Poland, Hungary, possibly Slovenia); and those (Bulgaria, Romania, Albania, the southern parts of former Yugoslavia) which have greater cultural and economic similarity to the south-east Mediterranean countries. In fact their characteristics are part East European, part Balkan and part Mediterranean. The third circle is formed by the southern and eastern rims of the Mediterranean, from Morocco along North Africa and up to Turkey. These are the semi-developed countries of the Preferential Interest Agreement who were hurt by the privileged access to northern European markets gained by the southern EU countries in the 1980s, but who are now being courted a little more seriously by the EU under the Euro-Med partnership established at the Barcelona conference in 1995. Beyond the Mediterranean rim countries lies the fourth group – the Lomé countries of Africa, the Caribbean and the Pacific.

From the perspective of the EU and this book, the intriguing comparison remains that between the second and third circles, and between the eastern and the Mediterranean frontiers. It is sometimes cynically observed that the EU will expand eastwards as far as the German taxpayer will allow and southwards as far as immigrants can be kept out. The southward expansion of the EU in the 1980s came at a time when emigration from Greece, Spain and Portugal had already died away and was giving way to immigration. The migratory threat from the east, widely seen as problematic in the late 1980s and early 1990s, has largely failed to materialize, partly because of the EU's skill in creating a cushion of 'buffer states' (Poland, Hungary, the

Czech Republic) to control the pressure of migrants in transit for Europe, so that the migration frontier has been displaced outwards.[43] Little of this nature has occurred with the southern Mediterranean countries, with the result that the migration frontier is sharply demarcated along the economic and demographic fault-lines described earlier in this chapter.

There is no doubt that the permeability of the EU's southern external border to migrants and other 'unwanted products' (arms, drugs, etc.) continues to give rise to concern amongst European policy makers. The southern European countries' position in the Schengen Group is compromised by this permeability, leading to particular problems with Italy and Greece which in addition to long sea boundaries have rugged land frontiers to monitor. Spain, meanwhile, has imposed a visa requirement on visitors from the Maghreb states. But it is difficult to foresee how tight control over clandestine migration can be achieved either practically or politically. Perhaps a greater concern for humanitarian issues – combating racism and facilitating short-term migration by quota systems and the granting of seasonal work permits for specific employment sectors – is the way forward as far as 'internal policy' is concerned; whilst external policy needs to go much further than the unrealistic attempt to stem migration at source by packages of aid, trade and development agreed at the Barcelona conference. Of course, managing international migration flows is not an easy task, but until a higher degree of trans-Mediterranean co-operation is reached on this and other issues, the southern flank of Europe will continue to be regarded as the 'hostile frontier'.

Notes

1. Braudel (1972).
2. See, for example, Branigan and Jarrett (1975); Ribeiro (1983).
3. Gillespie (1994); Kliot (1989).
4. Montanari and Cortese (1993).
5. Grenon and Batisse (1989), p. 1.
6. Rufin (1991).
7. King (1996).
8. UNDP (1996), p. iii.
9. Ibid., pp. 6–7.
10. See Sporton (1993).
11. King (1993).
12. See Grenon and Batisse (1989), pp. 46–57.
13. Livi-Bacci and Martuzzi Veronesi (1990).
14. Jones (1997).
15. Ibid.
16. Tsoukalis (1977).
17. Gillespie (1994), p. 1.

18. Anon (1995a).
19. Anon (1995b).
20. Anon (1995a); Jones (1997).
21. Anderson and Fenech (1994), p. 20.
22. Kliot (1997).
23. Joffé (1994).
24. Holmes (1995), p. 1.
25. Lesser (1996), p. 261.
26. Ibid., p. 262.
27. Anderson and Fenech (1994), pp. 14–15; and Lesser (1995), p. 22.
28. Kliot (1997).
29. Lipietz (1993), p. 501.
30. Rufin (1991).
31. Lipietz (1993), p. 509.
32. Amin (1989), pp. 90–4.
33. Sibley (1988).
34. Hall (1995).
35. Jess and Massey (1995), p. 168.
36. Fakiolas and King (1996).
37. Montanari and Cortese (1993).
38. King and Rybaczuk (1993).
39. King and Konjhodzic (1996), p. 58.
40. King (1996).
41. Anon (1995b).
42. Lipietz (1993).
43. Collinson (1996).

Bibliography

Amin, S. (1989) *Eurocentrism*, London: Zed Books.

Anderson, E. and Fenech, D. (1994), 'New Dimensions in Mediterranean Security', in R. Gillespie (ed.) *Mediterranean Politics*, London: Pinter, pp. 9–21.

Anon (1995a) 'A new crusade', *The Economist*, 2 December, pp. 27–8.

Anon (1995b) 'Club Med', *The Economist*, 2 September, p. 28.

Branigan, J. J. and Jarrett, H. R. (1975) *The Mediterranean Lands*. London: MacDonald and Evans.

Braudel, F. (1972) *The Mediterranean and the Mediterranean World in the Age of Philip II*, London: Fontana Collins.

Collinson, S. (1996) 'Visa Requirements, Carrier Sanctions, Safe Third Countries and Readmission: The Development of an Asylum Buffer Zone in Europe', *Transactions of the Institute of British Geographers*, **21**, pp. 76–90.

Fakiolas, R. and King, R. (1996) 'Emigration, Return, Immigration: a Review and Evaluation of Greece's Postwar Experience of International Migration', *International Journal of Population Geography*, **2**, pp. 171–90.

Gillespie, R. (1994) 'Introduction: Our Focus on the Mediterranean', in R. Gillespie (ed.) *Mediterranean Politics*, London: Pinter, pp. 1–6.

Grenon, M. and Batisse, M. (1989) *Futures for the Mediterranean Basin*, Oxford: Oxford University Press.

Hall, S. (1995) 'New Cultures for Old', in D. Massey and P. Jess (eds) *A Place in the World?*, Oxford: Oxford University Press, pp. 175–215.

Holmes, J. W. (1995) 'Introduction', in J. W. Holmes (ed.) *Maelstrom: The United States, Southern Europe, and the Challenges of the Mediterranean*, Cambridge, Mass.: World Peace Foundation, pp. 1–10.

Jess, P. and Massey, D. (1995) 'The Conceptualisation of Place', in D. Massey and P. Jess (eds) *A Place in the World?*, Oxford: Oxford University Press, pp. 45–86.

Joffé, G. (1994) 'The European Union and the Maghreb', in R. Gillespie (ed.) *Mediterranean Politics*, London: Pinter, pp. 22–45.

Jones, A. (1997) 'The European Union's Mediterranean Policy: from Pragmatism to Partnership', in R. King, L. Proudfoot and B. J. Smith (eds) *The Mediterranean: Environment and Society*, London: Arnold, pp. 155–63.

King, R. (1993) 'Italy Reaches Zero Population Growth', *Geography*, **78**, pp. 63–9.

King, R. (1996) 'Migration and Development in the Mediterranean Region', *Geography*, **81**, pp. 3–14.

King, R. and Konjhodzic, I. (1996) 'Labour, Employment and Migration in Southern Europe', in J. Van Oudenaren (ed.) *Employment, Economic Development and Migration in Southern Europe and the Maghreb*, Santa Monica, Calif.: RAND, pp. 7–106.

King, R. and Rybaczuk, K. (1993) 'Southern Europe and the International Division of Labour: from Emigration to Immigration', in R. King (ed.) *The New Geography of European Migrations*, London: Belhaven Press, pp. 175–206.

Kliot, N. (1989) 'Mediterranean Potential for Ethnic Conflict: Some Generalisations', *Tijdschrift voor Economische en Sociale Geografie*, **80**, pp. 147–63.

Kliot, N. (1997), 'Politics and Society in the Mediterranean Basin', in R. King, L. Proudfoot and B. J. Smith (eds) *The Mediterranean: Environment and Society*, London: Arnold, pp. 108–25.

Lesser, I. O. (1995) 'Growth and Change in Southern Europe', in J. W. Holmes (ed.) *Maelstrom: The United States, Southern Europe, and the Challenges of the Mediterranean*, Cambridge, Mass.: World Peace Foundation, pp. 11–28.

Lesser, I. O. (1996) 'Southern Europe and the Maghreb: US Interests and Policy Perspectives', in J. Van Oudenaren (ed.) *Employment, Economic Development and Migration in Southern Europe and the Maghreb*, Santa Monica, Calif.: RAND, pp. 261–74.

Lipietz, A. (1993) 'Social Europe, Legitimate Europe: the Inner and Outer Boundaries of Europe', *Society and Space*, **11**, pp. 501–12.

Livi-Bacci, M. and Martuzzi Veronesi, F. (1990) *Le Risorse Umane del Mediterraneo: Popolazione e Società al Crocevia tra Nord e Sud*, Bologna: Il Mulino.

Montanari, A. and Cortese, A. (1993) 'South to North Migration in a Mediterranean Perspective', in R. King (ed.) *Mass Migrations in Europe: the Legacy and the Future*, London: Belhaven Press, pp. 212–33.

Ribeiro, O. (1983) *Il Mediterraneo: Ambiente e Tradizione*, Milan: Mursia.

Rufin, S. C. (1991) *L'Empire et les Nouveaux Barbares*, Paris: Lattès.

Said, E. (1978) *Orientalism*, London: Routledge.

Sibley, D. (1988) 'Purification of Space', *Society and Space*, **6**, pp. 409–21.

Sporton, D. (1993) 'Fertility: the Lowest Level in the World', in D. Noin and R. Woods (eds) *The Changing Population of Europe*, Oxford: Basil Blackwell, pp. 49–61.

Tsoukalis, L. (1977) 'The EEC and the Mediterranean: is "Global" Policy a Misnomer?', *International Affairs*, **53**, pp. 422–38.

UNDP (1996) *Human Development Report 1996*, New York and Oxford: Oxford University Press.

Chapter 9

After Brobdingnag: Micro-states and their Future

TOM NAIRN

Introduction

This chapter deals with a hidden factor in the new 'empire of civil society':[1] the significance of scale. Assumptions about the average, typical or desirable nature of political scale have conditioned most thinking about borders and frontiers. The 'normal' borderline is that between two or more normal polities, almost automatically envisaged as having (among other things) a certain minimal scale. In theory also it has been taken for granted that the ideal-typical nation-state must be of a certain size, and plausible historical and sociological reasons have been advanced for this.[2]

Yet there are now many indications that this assumption no longer holds. The chief of these is the spread and relative success of extremely small states, the 'micro-states'. There was no place for such absurdities among the forbidding landscapes of Realist Theory: Singapore, Malta, Andorra (etc.) might be tolerated as eccentricities but had no real right to (so to speak) theoretic existence. Their significance for the World-Spirit remained that of joke, relic or awful warning.

However, any more developmental view of international relations must now try to redefine their significance. It is not an accident that the transnational or globalized economy (1960–present) has become such a successful habitat for them. 'Globalization' entails the formation of new frontiers, these among them – and (it can also be argued) in their way exemplary. The downside is their common repute as havens of sleaze, evasion, etc. Is this justified? Or should metropolitan and transnational *idéologues* look harder for the mote in their own eye?

There is fortunately no cure for micro-states. 'Average' nation-state scale was never in truth a by-product of either human, cultural or developmental needs alone. It also reflected the militarism, the balance-of-power and the rural-ethnic nationalism of a (mercifully) vanishing world. By contrast, micro-state and mini-state nationality-politics and borders seem likely to be primarily civic, treaty-dependent, multiculturalist and 'outward-looking'.

Scale and nationalism

Most serious theorizing about nationalism has gone on since the Second World War. No doubt inevitably – since inquirers were looking back on pre-war history – most of it has tended to assume a typical or average scale as applying to the units or entities it is concerned with. The matter has been endlessly debated, but I do not think a Cook's Tour of definitions is needed here. I will assume that most of the time the ideal-typical 'nation-state' or theorist's building-block has been 'something like France'.

This model naturally comprised many different assumptions about physical size, likely population, economic resources or development and culture. It will be perfectly well understood by everybody that France itself has always borne at best an uneasy half-resemblance to the shining type-model extrapolated from it. In fact, Benedict Anderson's famed dictum about 'imagined communities' applies notably to the template, as well as to the motley bunch of no-hope look-alikes, throw-backs and wannabees who nowadays crowd out the corridors of the UNO building. Gallic dream-merchants and tidy-minded foreign admirers alike have conspired to transform the peculiar French agglomeration into *l'hexagone*, a near-miraculous gift of nature and civility combined. Shifty, ambiguous reality counts for little on the plane of grand generalization and international-relations theory. Up there it is *la France* – the Platonic notion of Absolute Kingship, 1789, Jacobinism, *une certaine idée*, and so on – which has remained the forge, the holy matrix or womb of all modern state-formation. Hence it remains the most convenient way of summing up these assumptions for us today.

But they can be given an even more succinct summary, also worth mentioning. This is through a single word: 'viability'. *Viabilité, chances de succès*: the idea insinuates itself with suspect naturalness into practically every discussion of matters national or nationalist. The implication is that some specific package of minimal necessary conditions is a prequisite for modern statehood (or occasionally – when reality hints otherwise – *real* statehood). Unless these conditions are fulfilled, then a population or territory should have the decency to give up trying: they just have no chance of the real thing, nation-statehood. That is, *proper* and *effective* nation-statehood. One cannot be a respectable or non-token building block of the modern international order without the minimal equipment. And a certain size is demanded for possession and utilization of that equipment.

On the level of high theory, the point was put across most influentially by Ernest Gellner. It was a key feature of the 'modernization theory' of nationalism first expounded in *Thought and Change* (1964) and then elaborated in *Nations and Nationalism* (1983). In these canonical texts he maintained that post-eighteenth century political nationalism was in essence a world-wide forced response to the rapid or shock impact of West-European-led (or later, Atlantic-led) modernization. While there is no need to recapitulate it here, it should be remembered that in the famous theory, scale was awarded a crucial role above all for *cultural* reasons. This is the theme to which Gellner most unfailingly returned in all his subsequent writings. To respond to the shock of the new, a nation must redefine itself through language, art and education; but no village or merely rural region can accomplish this – it cannot create the institutions needed for the task, and so will be fated for assimilation or marginalization. A certain scale is required for fabrication and maintenance of the homogeneous high culture, or 'breathing-chamber', without which no population can re-invent itself as an aspiring nation-state. Hence the standard modern trajectory of nationalism to nation-state implies a minimal scale, the equivalent of four wheels and a propulsion-mechanism. Unless a country was 'like France' in such basic ways – intelligentsia, sufficiently ethnocentric urban or rural proletariat, germinating *bourgeoisie* – it could not be configured in the standard ways by the novel pressures of development. Deprived of this configuration, an effective ethno-nationalism would not arise, and the nationality in question would languish among the also-rans. Not without a certain *Schadenfreude*, Gellner noted in *Nations and Nationalism* that the also-rans were likely ways to outnumber the attainers.[3] The implacable threshold of scale would condemn them to the sidelines of dependency, minority status or absorption.

This was never a merely theoretical assumption. It was, for example, the practical rule of thumb employed by the League of Nations in the 1920s and 1930s. Monaco and Liechtenstein existed then as today, exasperating small pebbles in the otherwise perfectly fitting shoe of viability's conventional wisdom. But the League viewed them as unworthy nuisances, feudal vestiges or accidents (etc.) capable only of provoking conflicts among serious nation-states. They were not allowed membership. It was not until the 1960s and later that the vestiges and nuisances flooded into the League's successor, the United Nations.

Lilliput and Brobdingnag

Of course, the ideal-type threshold never applied in the other direction. Or not, at least, with the same rigour and conviction. Plainly there have been plenty of polities not 'like France' in the sense of being far too big, hetero-geneous, polyethnic and multicultural. The globe used to be covered with them, and until the 1980s its destiny seemed governed by the two super-bigs, the USA and the USSR. The 'pyramid' of a modern educational system

> provides the criterion for the minimum size for a viable political unit. No unit

too small to accommodate the pyramid can function properly. Units cannot be *smaller* than this. Constraints also operate which prevent them being too large, in various circumstances; but that is another issue ...[4]

Another issue indeed, to which he never returned with the same conviction, until the USSR foundered and Central-East Europe was liberated in the 1980s.[5]

In practice it was the same story. Plenty of Ruritanias were sidelined, but Megalomanias tended to be forgiven. The Wilsonian norm was de facto assymetrical: it had to be lived up to at one end (below the threshold) but could be exceeded with impunity in the other direction. The British, French and other empires were accepted like the Soviet Union, India, Indonesia and China, because of their power. And also because – then as today – many people thought that the national state was obviously declining, indeed doomed, and equally obviously could only be replaced by something bigger, rather than something smaller. Lilliputs, therefore, would soon be forgotten on the world's inevitable route to one or another Brobdingnag. Incidentally, if there was ever a time between (say) 1800 and the present moment when the nation-state has *not* been obviously declining, doomed, anachronistic and overripe for replacement I would be glad (though surprised) to hear about it.

Global and local

From the vantage-point of the 1990s, how is this world altered! Changed utterly, and in a sense rather encouraging to protagonists of smaller-scale sovereignty. This is not, incidentally, a question of small being necessarily beautiful, any more than gigantic was lovely before it. I believe that scale is a question of structure and functionality, and not of either ethics or aesthetics. Since the 1970s, in the twenty years which culminated with the end of the Cold War and the dissolution of the Soviet *imperium*, conditions changed in ways favourable to the existence, the prosperity and the proliferation of tiny states.

What were the shifts which brought about this transformation? A wry summary of some of them can be found in Eric Hobsbawm's recent history of the twentieth century, *The Age of Extremes*. He points out that though on the whole the world economy of the post-war boom period (1950–75) remained an 'international' or home-centred one in the old pre-war sense, there were significant signs of change in the later 1960s. It was at this time that a 'transnational' economic system began to emerge – 'a system of economic activities for which state territories and state frontiers are not the basic framework but merely complicating factors ...'.[6] This new system was characterized by transnational firms ('multinationals'), an accompanying international division of labour, and the rapid rise of offshore finance.[7] He goes on to single out the last factor, offshore money and financial movements, as perhaps the most important novelty of the era – the thing which provided the vital leverage for transnationalism to emerge and begin to seriously affect the world. But it was also the thing which most directly transformed the possibilities for very small states.

The huge Cold War expenditure abroad of the American government and the expansion of US-based companies into multinationals created the Eurodollar, a new de facto world currency replacing the Pound sterling. There was of course no state authority regulating Eurodollar transactions, the consequence being, in Hobsbawm's words, 'a vast, multiplying flood of unattached capital that washed round the globe from currency to currency looking for quick profits . . .'.[8] By the 1980s, a forced freedom from exchange controls had become general, assisted by the ideological wave of anti-state liberalization and privatization which marked the whole decade. Existing nation-state economies initially tried to protect themselves against its effects, but were forced to give up. The micro-economies of places like Monaco, Liechtenstein and Jersey, on the other hand, experienced prodigious and sustained development.

> Such units had been regarded as economic jokes, and indeed not real states at all . . . but in the Golden Age it became evident that they could flourish as well as, and sometimes better then, large national economies by providing services directly to the global economy. Hence the rise of new city states (Hong Kong, Singapore) a form of polity last seen to flourish in the Middle Ages . . .[9]

He focuses upon Asian examples here, but the same observation could be made over the same period about older units like the European micro-states (Andorra, Monaco, San Marino, Liechtenstein) as well as the greater number of new, mainly island states now scattered around the globe in the wake of decolonization.

Defining the micro

Who are they? One recent survey took as its criterion having a land area of less than 1000 square kilometres and a population of less than 500,000.[10] There is no obvious distinction between the micro and the very small, but if one follows this criterion then there are twenty-two in the ranks of today's United Nations. The majority of these are small islands in the Pacific and Indian Oceans or in the Caribbean, with one in the Mediterranean (Malta). Most of these were former British colonies. Unquestionably it has been the collapse of the British Empire which has bequeathed most micro-states to the New International Order or Disorder. French decolonization followed a different strategy of political incorporation, making similar tiny territories into *Départements* of the homeland. The British followed the contrary strategy of disincorporation: getting rid of them by making them independent, with or without bribes. The one case where that policy was not pursued, the Falkland Islands or Malvinas in the South Atlantic, brought unfortunate consequences in 1982.[11]

They are extremely tiny. In Europe, for example, the total population of all five micro-countries amounts to only 136,000 souls (the population of Bournemouth), occupying an area of 687 square kilometres (most of this being the uninhabited higher mountains of Andorra). These are figures too small to count as a percentage of the European Union's present total of 326 million inhabitants spread over two-and-a-quarter million square kilometres. It should be noted right away, though, that figures like 22 or 34 are no reliable indicator of micro-state reality, since a considerably larger number of small territories are 'self-governing' (notably self-governing in an economic and fiscal sense) without being members of the United Nations. The United Kingdom itself comprises a number of these, for example, like Jersey, Guernsey, Gibraltar and the Isle of Man – territories which so far have not sought the status of formal independence or sovereignty.[12] If these are taken into account, then there are over 60 such miniscule countries in today's political world.

Tiny states are jokes, rarely referred to in the metropolitan media except in terms of quaint happenings and uniforms – the equivalent of the 'feudal vestiges' or 'left-overs' theory mentioned earlier. The sole alternative to this seems to be the reprobates theory, which views them essentially as disgraceful and probably germ-laden fleas of the world order, about which (unless one has an awful lot of money) the less said the better: tax-havens, unseemly focuses of conspicuous or super-rich consumption (Monaco), or vulgar pustules of duty-free commerce (Andorra).

However deplorable such practices may be, the persistent tone of prejudice here should sound a warning bell. Hobsbawm's point about their place in the new economic system is in no way invalidated. Indeed, the contrary point can all too easily be made. Singapore did not cause the failure of Barings Bank; it was the operations of Barings (and a hundred other banks like it) which made Singapore what it has become in the multinational economic order of today. Liechtenstein's sleaze and secrecy stand as nothing set against the corruption of, for example, the Italian Republic since the 1950s, or the Spanish Socialist *régime* of Felipe Gonzalez in the early 1990s. In a period when sleaze has become universal, and public perception of large-scale political life and motivation has sunk to the low common denominator displayed practically every day on every Western newspaper front page, there is something exaggerated about so much outrage.

Anarchy and scale

But as anyone who becomes interested in this odd phenomenon quickly discovers, the plain fact is that most people dislike and despise tiny states. Or at least they do so publicly, or officially. A mixture of motives seems to be involved. These include loss of conviction about their own formerly great nation-states, disapproval of criminal money-laundering and understandable scandal at fat cats 'getting away with it' through mechanisms forbidden to ordinary citizens. But also (less admirably) a straightforward envy of places where, in a congenial climate, other people either pay no VAT and income tax at all, or far less than what we are used to. Publicly, it is outrageous that the residents of mere joke-countries should be so much better off than the nationals of serious, history-worthy states, including many an ex-Empire. Privately, the outraged often wish they could just move there. Officially the scapegoats must be castigated; unofficially, inquiries are made about residency criteria and the discrete setting-up of an offshore company.

Hobsbawm's history is sub-titled 'The Short Twentieth Century', but there is now another analysis which, I believe, carries understanding of the same point farther: Giovanni Arrighi's *The Long Twentieth Century: Money, Power, and the Origins of Our Times*. In his 'introduction', Arrighi refers back to what he calls 'the conventional view' of statehood and inter-state power 'consisting primarily of relative size, self-sufficiency and military forces'[13] – the same original, developmental model we see reflected in Gellner's theory. But, he underlines at once, this model has never been essential to capitalist economic development as such. Nor is it only since the 1970s that the two have diverged. It may be true that for a particular epoch development favoured the formation of larger competitive entities, the nation-state units of (approximately) 1750 to 1950. First-wave industrialization had to emancipate itself both from the confines of the city-state (where capitalism had always been at home) and also from the bureaucratic hierarchies of the ancient empire-states. This is why everywhere had, for a time, little option but to be, or try to be, 'something like France'.

But that time is over. If one over-identifies its political template with economic and social necessity *as such*, then one ends up misunderstanding the very large and the very small alike – for example:

> The capabilities of some Italian city-states over several centuries to keep at bay
> ... the great territorial powers of early-modern Europe would be as incomprehensible as the sudden collapse and disintegration of in the late 1980s and early 1990s of the largest, most self-sufficient, and second-greatest military power of our times: the USSR ...'[14]

After 1989, first-wave capitalist development reached an important boundary. The disintegration of the USSR was also the attainment of that limit, in the world-wide formation not of a traditional military-political empire but of something better described in Justin Rosenberg's recent phrase – 'the empire of civil society'.[15] This can in his view equally and more tellingly be called the final generalization of anarchy – not in a loose rhetorical or philosophical sense, but denoting the new circumstances of a global market-place, accompanied by political liberalization or democracy.

World-wide or 'globalized' anarchy in that sense denotes a system in principle unified along certain structural lines – unified and (for the first time) liberated from any prospect of 'empire' in the old political and colonial or territorial sense. Unified, in other words, but not of course united politically or territorially. On the contrary, tendentially more disunited than at any time since early modern Europe set up the provisional political template of modernity with the Treaty of Westphalia in 1648. Among the symptoms of the anarchy are that disconcerting reappearance of city- and micro-states Hobsbawm notices, and their troubling tendency to become more numerous, more economically important and to get themselves permanently established and recognized in the United Nations.[16] Picturesque theories like Alain Minc's *Le nouveau moyen age* (1994) have envisaged this as a sort of regression, or relapse into quasi-medieval conditions, but I suspect these lay too much stress on superficial similarities. They may also be influenced by the broader millennial and multinational industry of doom prediction.

On the other hand, non-doom predictors include writers like John Naisbitt. He starts off his recent *Global Paradox* (1994) with an evocation of the then newly independent Andorra (population 54,000, mainly non-Andorran residents) which suggested that 'the world's trends point overwhelmingly towards political independence and self-rule on the one hand, and the formation of economic alliances on the other'.[17] Andorra's population is less than half that of England's Isle of Wight (125,000, 381 square kms, capital Newport), which waited until 1996 to make its move in the same general direction.[18]

Implications of Lilliput

Soothsaying apart, what are likely to be the implications of Lilliputian revival and statehood for post-1989 boundaries and identities? 'What benefits, or what misfortunes to mankind may hereafter result from these events, no human wisdom can foresee', wrote Adam Smith in *The Wealth of Nations*, in 1776. He was discussing the tendency of capitalism to '. . . unite, in some measure, the most distant parts of the world, by enabling them to relieve one another's wants, to increase one another's enjoyments, and to encourage one another's industry', a tendency beneficial on the whole although qualified by appalling misfortunes and oppressions. He was discussing, well ... 'globalization' (proof, were any needed, that familiarity with *Star Trek* and World System Theory is not required for this task). The trouble, he thought, was 'the superiority of force' which then happened to be on the side of the Europeans, and which made development not just Eurocentric but consistently brutal and antagonistic. Only once greater equality was established and guaranteed everywhere could that be righted. That – a post-imperial climate of international affairs, as it were – 'can alone overawe the injustice of independent nations into some sort of respect for the rights of one another'.[19]

Some doom predictors believe that this will never happen. Irrepressible optimists like me think it may be happening already. And one of the arguments for the sanguine view must be the phenomenon of the thriving and proliferating micro-state. One thing which goes on irritating theorists, especially grand-metropolitan theorists, is that fleas are not mammals. They are not real nations. They are indeed pretend-nations, miniature simulacra of true nation-statehood, that noble condition for which masses have struggled, armies have fought and mighty empires have crashed to the ground. Not being real countries, they cannot be really independent; nor do they seriously pretend to be other than dependent, itself an almost sinful form of *lèse-majesté*. Hobsbawm is particularly irked by it, and argues that no one should be taken in: 'The most convenient world for multinational giants is one populated by dwarf states or no states at all', he snarls in conclusion.[20] Such independence cannot therefore be worth having, and all that Singapore or Andorra are doing is providing a kind of political alibi for the new ogres of economic imperialism.

I suggested earlier that moral thunderings about micro-state sleaze are themselves suspect, and in a sleazily liberated world can be easily stood on their head. Is not the same true here? Of course today's tiny and city-states are not 'real' in the sense most characteristic of the Age of Nationalism. They never corresponded to being anything like France, or to the theoretical model promulgated in theories of modernizing nationalism like Gellner's. In one sense banal, the accusation is also misleading in that (like the tax-haven sermons) it ignores the now quite general circumstances of dependence, vulnerability and objectively diminishing sovereignty. If these are pretend-polities, so increasingly are we all. In 1992, the United Kingdom (original founder-member of EMS, template *extraordinaire*, all-purpose model and leader) was forced to devalue its currency by an international speculative movement orchestrated by the financier George Soros, through his *Quantum Fund*. Though managed from New York, his off-shore fund happened to be head-quartered in the Caribbean micro-state of Curaçao.

Micro-statehood is 'synthetic', runs the complaint; since they have no armies or missiles, tiny states rely upon artificial conventions of sovereignty maintained by international law and treaties. They depend mainly upon (in Smith's words) 'some sort of respect for the rights of one another' continuing to prevail over the 'injustice of independent nations' larger than themselves. The five original European examples all count upon special treaties with the European Union, by which, without being members, they retain certain privileges and are (in effect) granted absolution for their assorted fiscal or commercial sins, and permitted to carry on business. So do the United Kingdom's semi-states of Jersey, Guernsey and the Isle of Man. Their frontier with Europe is in this sense essentially an odd set of legal provisions which includes them out, as it were – the very opposite of that tradition of superiority of force which Smith denounced.

What I do not see is what is *in any way* deplorable about artifice of this kind. On the contrary, the recognition and tolerance of micro- and mini-states is an expression of civilized grandeur, and (the same thing) of emancipation from the savage Armageddon climate which disfigured the later Age of Great-Power Nationalism. The 'independence' which it recognizes and fosters is not, *of course*, in the traditional style hallowed by ethno-nationalist ritual, romantic culture, the cult of war-memorials and *le sol sacré*. These things too (we should recall) were part of *la France*'s template. However, it is both a theoretical and (more seriously) a political mistake to think they are inseparable from national identity or character, nationality-politics or the general – and growing – need for self-government.

The more interdependent countries become socio-economically, the more political 'independence' is likely to be needed. Such independence can only be civic, negotiated and inseparable from both internal democracy and external system-support. Some new nationalist movements like those in Wales, Scotland, Catalonia, Sardinia (and many others) now envisage 'independence-in-Europe'. Quebec's *souveraineté-association* also was a way of having independence while trying to remain in some way linked to the Canadian Federation. Denounced by old-style sovereignty-addicts as 'having it both ways', these formulae may also be seen as precursors of a new era where all countries and populations (including the USA and China) 'have it' – and in a sense *have* to have it – in different ways at once. Arrighi's point is that they always did: autarchy and ethnic nationalism were myths based on legends, while interdependence and civic nationalism are the long-haul parameters.

As they re-emerge from the warlike universe of 1789–1989, we can see how scale has changed in meaning. When peasants had to be conscripted into armies, factories, 'national' languages and cultures, in order to be something like France, of course a minimal socio-economic and territorial scale was unavoidable. First-round or primitive industrialization demanded security, as well as a Gellnerian 'breathing-chamber' or 'lock' system.[21] Construction of such a system implied a de facto minimum scale of operation, more easily realizable in some parts of the human culture palimpsest than others. This first-phase building of nationalism (ethno-linguistic, romantic and Janus-faced) steered modernization through the much greater dangers of imperialist take-over and colonization.

However, that was also a once-only transition process, and approximately completed in the 1980s. During it, borders and frontiers acquired a deadly, life-or-death meaning. Too much depended upon them. They were 'inviolable' – sacred emblems of a precious, fought-over and sustainable difference of rights and privileges supposed defensible only upon a certain scale. Polities can in fact be of any dimension, as city-states have demonstrated throughout recorded history. But the post-eighteenth-century tribe-nation mystique demanded a certain amplitude or convergence of *ethnies*: the gateposts of modernity, as it were, passage through which alone could render industrialization tolerable.[22] On the other hand, once that passage is accomplished, the gateposts turn into monuments. And so does the scale-factor which for a time they imposed upon development.

On the Saarland-Lorraine border, one of the most contested old nation-state lines, a project was launched in 1991 to create a new frontier of sculptural forms – the bournes of an age to come, as it were, recognizing at once the futility of former defences and the persisting need for markers of a different kind. Better than any words, its strange scatter of stone shapes across the undulating French-German field-landscape represents this will for the new. Over that horizon lies the unparochial localism which, for the first time, globalization is making possible.

Notes

1. The name is Justin Rosenberg's, from his book of the same name (see below, note 15).
2. Most famously in the work of the late Ernest Gellner, including his magisterial *Nations and Nationalism* (1983, Oxford: Oxford University Press).
3. See Gellner, *Nations and Nationalism*, Chapter 4, 'A Note on the Weakness of Nationalism'.
4. Gellner, *Nations and Nationalism*, pp. 43–50.
5. One of his last writings was a puzzled meditation on the reasons for the sudden and complete collapse of the Russian-Soviet imperium, 'The Rest of History', *Prospect* (May 1996).
6. Eric Hobsbawm (1994) *The Age of Extremes: The Short Twentieth Century*, London: Abacus, p. 277.
7. The term 'off-shore' entered the general vocabulary of most languages around this time. It can be found, for example, in the *Petit Larousse* dictionary, listed as reluctant French ('anglicisme déconseillé') rather than in the pink pages at the

middle of the book as a *locution étrangère*. It meant the practice – originally
marginal and apparently of minor importance – of registering the legal seat of
businesses in some tiny and fiscally generous territory which allowed firms to
avoid taxes and other constraints imposed by their own countries.

8. Eric Hobsbawm (1994), p. 286.

9. Ibid.

10. *L'evenement du jeudi* (Paris, 28 juillet–3 août 1994), 'Les 22 plus petits états du
monde', Michel Chabot. A more academic attempt at defining the terrain has
been made by Laurent Adam in 'Le concept de micro-Etat: Etat lilliputien ou
parodies d'Etats?' in *Revue Internationale de Politique Comparée*, Vol. 2, No. 3
(1995). Adam's definition is more ample than Chabot's: 'Le nombre d'habitants
inférieur ou égal à 1 million; la superficie inférieure à 6000 km² . . .' (ibid.,
p. 587). This gives a total of 34 micro-states in the UNO list. The most useful and
comprehensive overall survey may be that published in Liechtenstein: Arno
Waschkuhn's *Der Kleinstaat in globaler Sicht* (Sophistocat Press, Ruggel & Mün-
chen, 1991). I am grateful to Gwo-Jinn Hwang for drawing my attention to this
book, as part of his new study of European micro-states.

11. Those who suggested at the time the sole way out might be artificial statehood
for the islanders were derided. Was it not obvious that independence for just
2000 people on a few barren rocks would always be . . . wholly unviable? Thus
Royal-British empiricism faltered fatally at one of its last outposts. 'God who
made Thee micro, make Thee microer yet . . .' – but not (or not yet) in Port
Stanley or Gibraltar.

12. The former British colony of Bahamas recently held a referendum on statehood
which returned a majority for retaining the links with Great Britain.

13. Giovanni Arrighi (1994) *The Long Twentieth Century: Money, Power, and the
Origins of Our Times*, London: Verso.

14. *Ibid.*, pp. 15–19.

15. Justin Rosenberg (1994) *The Empire of Civil Society: a Critique of the Realist Theory
of International Relations*, London: Verso, especially Chapter 5, section 6, 'Karl
Marx's Theory of Anarchy'.

16. Six have accomplished this since 1990: Liechtenstein (1990), The Marshall
Islands and Micronesia (1991), San Marino (1992), Monaco and Andorra
(1993). It is especially noticeable that they may now beat larger competitors to
the mark. No Andorran, for example, will miss an opportunity to emphasize,
especially in the presence of someone from Barcelona, that the Catalan language
and culture first resounded in the international arena through *their* admission,
and not that of Catalonia itself.

17. John Naisbitt (1994) *Global Paradox: The Bigger the World Economy, the More
Powerful its Smallest Players*, London: Nicolas Brealey.

18. I wrote an account of the episode in the *Demos Quarterly Bulletin* (No. 9, 1996),
under the title 'The Incredible Shrinking State'. The island is effectively seeking
free-port and tax-haven status comparable to that enjoyed by the Isle of Man and
Jersey or Guernsey.

19. Adam Smith (1776) *The Wealth of Nations*, as quoted in Arrighi, op. cit., p. 19.

20. Eric Hobsbawm (1994), p. 281.

21. The metaphor of canal or harbour locks comes from Gellner's original outline of

modernization theory, at the end of his famous essay 'Nationalism', in *Thought and Change*, London: Weidenfeld and Nicolson, 1964.

22. The most distinguished meditation on these themes is Martin Thom's 1995 study, *Republics, Nations and Tribes*, London: Verso Books.

Chapter 10

Frontiers and Security in the European Union: The Illusion of Migration Control

DIDIER BIGO

Frontiers and the cross-border flow of people within the European Union

The phenomenon of transnational flows of people has a long history. But these flows seemed to accelerate after the Second World War, and an increasing number of players other than states appeared on the international scene. This could be described as a *'retournement du monde'*[1] (a global turning point). The spread of capitalist economies throughout the world, the emergence of economic agents with many transfrontier links since the 1950s/1960s, called into question the capacity of the states to implement different, or genuinely independent, economic policies. The economic field no longer coincides with nations or their frontiers. The models of Adam Smith and Keynes view the economies of nations as units, but wealth is now increasingly disconnected from nations and circulates within cross-border networks – either at the level of production (international firms) or, and especially, of finance (international banks, speculation . . .). Many of them are composed of multiple and autonomous players seeking the most profitable short-term alliance and not absolute power.[2] But wealth has not, as a consequence, been more evenly spread. Certain zones (the European Union, NAFTA, Japan and the new industrialized countries) concentrate capital and technology and attract labour – men and women are more mobile than the other factors of production.[3]

Cross-border flow of people is associated with economic globalization, with the process of the international division of labour, and with the economic cycle, although the demand for labour is present even during periods of crisis. But many other factors have to be taken into account – differential demographic development, internal migration in Third World countries and, especially in the countries which border the European Union or the United States, the effects of the diffusion of the 'Western model', linguistic policies affecting the major languages (English, French, German), the reinforcement of images through television, the explosion of the number of television channels via satellite, the promotion of the model of 'market democracy' to the countries which border the big blocs, the desire to escape authoritarian state control structures, and violent conflicts.[4] Migration increased rapidly during the 1960s and, insofar as it involved cross-border movements of people (immigrants, refugees, transfrontier workers and especially tourists), concerned the states; migrations were scarcely affected by the crisis in the mid-1970s. The problem of controlling large numbers of people at the frontiers was posed for states with prosperous economies, governed by the rule of law.

Controlling people at frontiers is a basic function which determines most cross-border practices. It involves the founding myths of the nation-states. Whether it is Hobbes' security pact, Locke's social contract, the monopoly of legitimate use of force in a given territory, to use Weber's formula, the different national or nationalist tendencies to present plural identities as if they were homogeneous and unique, these myths are used to justify the authority of the state.[5] Frontiers of states became territorial based codes of obedience in a binary form – one against the other, ones to be protected and ones to be mistrusted, friends and enemies. For a long time the enemy – the 'natural' enemy – was the territorial neighbour; the frontier served as a protection against the neighbour. But aerial warfare, the disappearance of any distinction between the military frontline and civilian zones and the development of atomic weapons destroyed this notion.[6] European integration and the Atlantic Alliance have, in the case of France, driven the enemy back from the immediate frontiers. The frontier was no longer a barrier, it could become a junction, a place for exchange. Economic growth, linked to the reduction of the barrier function of the frontier, has been diminishing the plausibility of the security argument concerning the protective function of frontiers.

Malcolm Anderson and Michel Foucher have explained in detail the profound transformation in conceptions of the frontier. As they contend, the frontier is never 'natural' or a matter of physical geography. It is always a political process – an institution defining difference with the outside world and attempting, by influencing mentalities, to homogenize the diverse population inside the frontier. It is therefore a political 'technology' which records the balance of power at a particular time in space.[7] From this starting point this chapter seeks to explain the remaining ambiguities in conceptions of frontiers, some of which are involved in discourse about immigration and coloured foreigners, relating to the old myths still present in memories and using the 'reflex reactions' of fear of closure, of tensions and of enmity. It seeks to analyse the political strategies involving increasing the salience of frontier controls. The symbolism of the frontier as a protection against danger still has appeal, and is frequently used in speeches on security. One can still mobilize populations with the spectre of the 'country in danger'. But border practices have changed irrevocably and, as Georges Tapinos has pointed out, this change is what will mainly be at stake in the next century.[8]

Immigration, as well as tourism, cross-border circulation and all kinds of nomadic behaviour, undermines classic conceptions of state capacity to govern. Contradictions between liberal, economic, political and security logic become apparent. Free movement of people cannot be managed in the same way as goods and services. Security and economic priorities do not coincide. Contradictions are highlighted by, on the one hand, a discourse about 'immigration control' while, on the other hand, by the practical strengthening of the bonds between the countries of the European Union. The national state seems destined to change – Europe looks like a confused attempt to find a post-national, post-state form where transnational flows correspond to (even anticipate) a transnationalization of the bureaucratic systems of control. The relationship between security and frontier is not a matter exclusively of immigration controls but must take into account all cross-border flows of people and also relate to power relationships and to the logic of control of the population of the European Union.

Europeanization and its consequences for frontier control

Immigration is often regarded as synonymous with the cross-border flow of people, but some phenomena creating these flows are often forgotten, in particular tourism. In the 1960s, immigrants were thought of as workers who would return home – 'long-term tourists' beneficial to the economy, working in a country where they were welcome, although brought up in another. With economic crisis, a new discourse on immigration emerged. One image – the single migrant worker – is replaced by another – the permanently resident immigrant family. In France, republican values were immediately contrasted with Islamic values, and marginality, immigration and crime were linked, and a negative relationship assumed between immigration, unemployment, family and social benefits. The immigrant, considered beneficial in a period of economic growth, acquires a negative image during economic recession, ruining the welfare state by fraudulently claiming social security and unemployment benefits. Clandestine workers, entering the country illegally or staying longer than allowed, breaking the law without regard to the consequences of their actions. The image of the immigrant merges into that of the unemployed, the thief, the smuggler and the criminal – an image used by parties of the extreme right.[9] The immigrant is the one who is different, who has other customs. White Americans disturb less than the Englishman of Pakistani origin or the French citizen of North

African origin. The debate on immigration becomes 'taboo' because public figures are afraid to discuss it for fear of helping to promote the views of the *Front National*.

Since other countries, such as Germany and the Netherlands, have relatively similar problems, it is tempting to transfer the question of immigration to the European Union.[10] But immigration policies and nationality entitlement cannot easily be standardized; they remain different because political interests as well as countries of origins of the migrants are too diverse. However, with the negotiation of bilateral agreements (commencing in the 1980s), the establishment of an ad hoc European immigration group (1986), the signing of the Schengen Agreements (1985 and 1990), with the Palma document (1989) and the Dublin Convention (1990), with the Third Pillar of the Maastricht Treaty (1992), the countries of the European Union embarked on a course of transforming immigration from a political question to a technical one, by presenting it as a matter of security.[11] The reduction of frontiers as barriers (tariff/customs barriers) and the principle of free movement was directed towards a dynamic opening up of European societies.[12] In the event, due to world recession, all the rich countries tended to take a tough stance on migration of people, contrary to the Union's aspiration to encourage the free movement of people, initially for European nationals but with an inevitable widening to others, in the absence of practical and legitimate criteria for discrimination. The White Paper of the European Commission on the Single Act had as its theme the abolishment of barriers, and created anxieties among those whose jobs depend upon controls at frontiers (for example, customs officers and border police). If the part of the White Paper on free movement was generous (and in the spirit of the founders of the Union), it also explained that protection against drugs and terrorism necessitated new forms of control. The distinction between internal and external frontiers of the Union was born.[13] Differences over this distinction continue, in particular between continental Europe and the British; the latter refuse to lose the right to 'insularity' and assert that they will maintain control at frontiers, as 'an essential element in the fight against crime and immigration'.[14] Among the Schengen countries, varying interpretations of the agreements make for significant differences in frontier controls, exemplified by the situation in the Netherlands and Italy.[15] Change in the global situation in the 1990s have caused further problems for the straightforward and generous idea of free internal movement with the controls pushed back to the external frontiers of the Union. Now, with the envisaged enlargement of the EU, it is uncertain which countries remain outside the Union and for how long is not fixed.[16]

Uncertain European frontiers

The territorial framework of the European Union is not stable. On the contrary, it is undergoing a profound transformation. Union members are not agreed on policy, not even in the harmonization of policy on frontiers and frontier control techniques. Agreement has not been reached on resident foreigners although 80 per cent of 'illegal' immigration is due to expiry of tourist visas, while only 20 per cent result from illegal frontier crossing.[17] Agreements are exclusively concerned with short-stay visitors and even on this issue the British are firmly committed to a different position. Denmark was in an uncomfortable position between the Union and the Nordic Union in the Europe of the Twelve. The expansion of EU members to 15 in 1995 introduced another complicated situation by isolating Norway as a non-member of the European Union but still within the Nordic free movement zone. The EU expansion projects favoured by the Germans and others in the IGC pose even more questions: membership of Poland, Slovenia and the Czech Republic would push back the EU frontiers a long way east. Internal migration problems would become as great as external ones. Would Schengen be 'an EU internal security cordon between the most developed countries and the others'?[18] Paul Masson in his last report envisages this role much more than the introduction of European citizenship and free movement.[19]

The instability of the framework, of the Union's exterior, is not simply geographic, it is more and more the instability of 'different Europes', differentiated according to policing areas. The single currency would not include the same countries as defence which would again be different to the 'home affairs' Europe. If there are as many Europes as pillars, this makes the task of defining frontiers very difficult. The maps of these different Europes are analogous to the maps of the Holy Roman Empire, with micro-states and overlapping jurisdictions.[20]

The geography of free movement will not follow the Union boundaries, and state-controlled territoriality cannot be managed as before.[21] To understand the relationship between flows of people and territories, we must return to concrete practices and the thinking which informs them. What is at stake is inherently symbolic and involves our conception of the state. The European Union is built on fears of imaginary enemies. It has analogies with processes of state formation; and states built on strong identities, on conflict and enmities play a part in building the European Union. This explains the heated controversies over immigration control (while there are none on tourism or cross-border activity), and it explains the distinction between community and non-community.[22] Above all, it explains a focus on erroneous assumptions, such as that it would be possible, if the political will existed, to control systematically foreigners at frontiers while allowing the citizens of the Union to cross freely.

The rhetoric of 'sieve' Europe and 'fortress' Europe

The critics of the Schengen Agreement and of the Third Pillar of Maastricht, quickly raised the question of the security deficit created by the abolition of internal frontiers. After their implementation, they present these texts as 'compensatory measures', to 'make good the security deficit'. The content of public debate is conditioned by the way its terms are set: in this case, the terms took the form of an artificial confrontation between those who found the content of the agreements insufficient (sieve Europe) and those who found them excessive (fortress Europe).

Sieve Europe

By reading only the main article in Schengen on free movement (article 2.1) without reading the other articles, some of those on the right accused Schengen and Maastricht of increasing the risk of international criminality and immigration. In France, Pierre Mazeaud emphasized in the National Assembly the risks of an influx of immigrants from Central Europe and Africa by, for example, taking advantage of the breach in Schengen external frontiers created by the German-Austrian agreement allowing East Europeans to enter Germany and from there to go to other EU countries. Immigration from the Maghreb could happen, Mazeaud alleged, through Italy or Spain whose lax frontier controls are notorious. Europe will be destabilized, argued François d'Aubert, by the infiltration of the Italian Mafia into France; drugs will come from the Netherlands, argued Gérard Larcher; Turkish immigrants, asserted Charles Pasqua, a theme later taken up by Jean-Louis Debré, driven out from Germany by the arrival of the Eastern immigrants, will invade France. Based upon forecasts by Lesourne and Chesnais of potential immigration from the Maghreb countries into Europe in the next millennium,[23] some Members of Parliament now in government spoke of 'an immigration explosion from the Maghreb countries'.[24]

Fears had already been mentioned in 1990 by Paul Masson, President of the Schengen Control Commission in the Senate: 'Each weakness, each act of clumsiness in implementing the agreements will lead, I am sure, to many French people contesting the idea of Europe itself which would become for them a Europe of insecurity and dubious activities.'[25] Xavier de Villepin and Masson particularly emphasized both the limited but recurrent problems of refugees from conflicts such as Yugoslavia and the necessity for 'burden sharing'. They noted: 'It is likely that the abolition of controls at internal frontiers will be interpreted as a signal to all the poverty-stricken of the world, in particular those in the South and the East' – reviving the spectre of an immigration invasion and, in a manner repeated by some members of the

European Parliament, justifying their position by the necessity of firmness now in order to avoid the need to be repressive later. They ended their plea: 'neither fortress Europe nor sieve Europe, the Europe of the Schengen Agreement must be a Europe of freedom because it must be a Europe of firmness'.

More extreme critics consider that free movement itself must be abolished and that control must be restored everywhere 'by keeping the frontiers'[26] and argue in favour of tougher entry conditions (visas in the name of prevention of terrorism, and the right of refusal without explanation, the computerization of consulates, the harmonization of Schengen procedures, penalties on airline companies), as well as frontier police and customs officials and the establishment of a frontier zone (a zone of 20 km as in Schengen with joint police stations). Since people are still getting in despite current precautions, it is argued that illegal immigrants already in the country must be located, marriages of convenience prevented, people pretending to be students or tourists identified and search and detection of illegal immigrants facilitated by checks in schools, social security offices, workplaces and places of residences. Law enforcement should be as strict as possible to deter new immigrants. Some consider the fight against illegal immigration is the necessary condition for allowing legal immigration.

New legislation has created a category of unauthorized persons who are not strictly illegal immigrants and who have been living in France for many years. The legal position of persons of foreign origin thus becomes a matter of permanent suspicion. Immigration control has become analogous with enemy infiltration, against which we have to be protected by systematic control at frontiers, by a sort of electronic Maginot line. Other Europeans can perhaps associate with this strategy, although even this association is controversial. A further step is taken almost surreptitiously to identify populations who 'cannot be assimilated' because of different religion (Islam), different morals (polygamy) or different ways of life. The issue becomes less the crossing of physical frontiers and more the transgression of symbolic frontiers imperilling the identity of the country.[27] The drive to close frontiers closes minds; criminalizes immigrants and makes immigration a problem of security.

This form of discourse is not confined to the Right in France – it closely corresponds to the discourse of German, Belgian and even Dutch conservatives. It also influences the French Left which, seeking legitimacy, does not wish to appear lax. Political milieus throughout the EU share a belief that controls at external frontiers must be seriously reinforced, and that the cooperation between frontier police forces is necessary to avoid the growth in crime and to support the policy of 'controlling immigration'. This consensus can be explained by the fears of the professionals of internal security and their influence as a lobby. Different agencies controlling the frontiers were particularly concerned by the speeches of pro-Europeans about the implications of free movement because their jobs were threatened. Customs officers and the officers of the border police in France informed the media about the dangers of opening frontiers. They produced statistics to prove that drug seizure at internal frontiers was substantial, and argued that all their efforts would count for nothing if frontier checkpoints were removed. European police unions warned about the worsening situation in the USSR, the development of East European 'mafias' and the new dangers in the post-bipolar situation. We have analysed elsewhere why and how an ideology widely shared by security forces across Europe has been disseminated: the ideology of a security continuum linking crime, immigration and asylum as a single threat called 'the Southern threat', supplemented by the theme 'international disorder', the proliferation of 'grey zones' and the 'clash of civilizations'.[28] An ideology of fortress Europe has been invented. This is no more 'real' than its opposite: sieve Europe.

Fear of reinforced controls: 'fortress Europe'

Humanitarian associations rejected the idea of sieve Europe since they saw reinforced controls and new levels of policing as involving a harmonization by the Fifteen at the most repressive end of the spectrum.[29] In France, Amnesty, the League of Human Rights, France Terre d'Asile (pro-asylum organization) criticize Schengen, stressing the risks of attacking the right of asylum and providing reminders that this is contrary to the Geneva Convention. These associations managed *in extremis* to include the right of the High Commission of Refugees to examine the articles concerning asylum. Although they denounced risks of 'compensatory measures', and accused Europe of withdrawing into itself, forgetting its duty to asylum seekers, they accepted compensatory measures as 'a necessary evil'. Their fight was a rearguard action to introduce 'safety clauses' of a judicial nature for countries where, in the name of effectiveness and speed, governments relied upon administrative procedures for dealing with asylum cases. The *Mouvement contre le Racisme, l'Antisémitisme et pour l'Amitié entre les Peuples* (anti-racist and peace movement) as well as the *Groupe d'Information et de Soutien aux Travailleurs Immigrés* (information and support group for the immigrant workers) went further and argued that the logic of Schengen is anti-democratic. They are among the few in France to make the link between the control of foreigners at frontiers and the situation of immigrants already in the country, seeing a possible drift from control to racial prejudice. But they have a small audience and give the impression that they are swimming against the tide; since the trend towards closure is unanimous at the level of the Fifteen, they seem to regard it as 'inescapable'.

In the other countries, humanitarian associations face the same difficulties.[30] Statewatch in England published a disturbing report on the state of civil liberties in Europe. The debate on 'fortress Europe' went further in the Scandinavian countries and included arguments to the effect that the opening of internal frontiers modified democratic practices by strengthening controls at external frontiers, by reinforcing controls over foreign populations already in the European Union, by increasing identity checks to locate illegal immigrants and by toughening the conditions to obtain asylum.[31] There was also, in France and in other European countries, fear of a revival of racism, of the rebirth of fascism and the extreme right. This fear overestimates the danger to freedom, forgetting that, statistically, few of the rejected asylum seekers are in the event escorted back to the frontier; overburdened police officers have the feeling it is they who are swimming against the tide. Tough policies recommended in the reports are not put into practice.

To mobilize their members, the civil liberties associations need a simple message and a popular slogan – in this case 'fortress Europe'. The image is plausible: a new wall is being built, this time created by the democracies defending economic privileges and refusing to consider the smallest increase in equity on a world scale.[32] The image of fortress Europe has obvious allusions: a new Cold War, militarization of controls, inhumanity, and so on. Could it mean that Europe is becoming racist and totalitarian? Could there be a police and army conspiracy against human rights? Agreements have been signed secretly – does that not point to such a plot? This kind of speculation leads to conclusions that there is a move in security due to plotting by police officers, and that there indeed is a plot – not a fifth column of immigrants – but the machinations of black reactionary forces. Some associations such as Amnesty are careful not to adopt this caricature, but others, often immigrant defence associations, appeal to the media believing that public opinion will react when it is made aware of this.

Although perhaps isolated, these associations are not alone in adopting the discourse on fortress Europe. The European Parliament accepted aspects of it, in particular in its resolution of June 1987. It subsequently became less polemical and complained only about 'lack of democracy', arguing primarily for reinforcing its own powers rather than for a change of policy. Governments of the third countries (especially in the Maghreb and Central Europe) interpreted as directed against them the Schengen and Maastricht tightening of police controls on immigrants and refugees. They pressured the governments of the Twelve into contorted statements on emphasizing international co-operation with countries bordering the EU. Countries of origin of immigrants initiated discussions with the EU on economic development to remove causes of migration as well as to support their efforts to eradicate drug transit and production. At the rhetorical level, the European Council made repeated statements on the risks of the creation of fortress Europe. Some officials consider the image of Europe abroad is already affected by anti-immigration measures and by the statements about 'immigration invasion'. This offends neighbouring countries and the other nationalities living within the whole European Union and reinforces their belief that Europeans refuse to accept the fact that resident non-Europeans are different.

Since the discourse on fortress Europe tends to be voiced only by those without power, it becomes propaganda, a way of mobilizing activists rather than a description of social practices. It is a reaction to the discourse on sieve Europe and shares a large part of its assumptions. Fortress Europe is present in the political arena only inasmuch as it is used occasionally by political parties who all want to give 'proof of their firmness', of their responsiveness to an alleged citizens' 'demand for security'. This discourse is not actually without substance because it reflects a general trend of government policies which, in the name of immigration control and burden-sharing between countries, increase legal obstacles to entering the territory of the European Union. This trend is often regarded as having gone too far, even more so if government rhetoric is taken to be a true reflection of practice.

A false prescription: maintenance of systematic frontier controls

The discourse of 'firmness' and the discourse of 'anti-racism' reflected a belief that governments can close the frontiers if they so decide or, at a minimum, they can control the foreigners entering their territory. There is an additional assumption that the legal rules have a practical application and thus change significantly cross-border flows. But one can legitimately ask two questions. Does anyone have a clear picture of cross-border flow of persons and the methods of control? What are the practical implications of tight control of foreign populations for economic prosperity and free movement of goods and services, civil liberties and liberal values?

Demands are made that the law be rigorously enforced even though the laws are based on erroneous assumptions about social practices. There is a desire to make reality conform to political objectives, and hopes are expressed that new governments, unlike old ones, are going to 'halt crime', 'declare war on drugs', 'eradicate terrorism', 'control immigration'. In respect of frontier controls, politicians are not ready to accept publicly that governmental autonomy is reduced by economic interdependence (bolstered by speeches on deregulation and the benefits of international trade). They cling to the dogma of sovereignty and turn security policies into a central political issue to regain some credibility. The immigrant becomes a political opponent of all security professionals who feel personally insulted that he or she has managed to cross the frontier. But even by co-ordinating their practices, harmonizing policies (bilateral agreements of police co-operation and readmission agreements between the main countries of the Union, the Maghreb, Central and Eastern Europe, the multilateral agreements of Schengen, the Trevi action programme, the complicated architecture of the Third Pillar of Maastricht), they are not able to control immigrants. This is because the public policies with regard to security and immigration are 'symbolic'. They are declarations which have dissuasive effects but they are not a practical programme of controlling the frontiers. The belief that controls are applicable simply because they are said to be, inhibits an understanding of the actual practices of control. A different perspective is required – we should 'consider these practices of borders control rationally. On what calculations and strategies are they based?'[33]

Practices at frontiers

The annual number of crossings by land, air and sea at external frontiers, delimited by the agreements of Schengen, including France, Germany, the Benelux countries, Spain and Portugal (still excluding Austria and Italy), totals about 1.7 billion crossings; 864 million cross the German frontiers alone. Much of this is local cross-border traffic, but clearly very strong economic and human links exist between the Schengen countries and their neighbours. Even temporary closure of the frontiers is unrealistic. Annual crossings of internal frontiers of the Schengen area number about 1.2 billion. Movement in and out of French territory totals 291 million each year – by air (600 airports), by sea (4720 km of coasts) and by road (2490 kilometres). Three-quarters, in the region of 230 million crossings, are made at the land frontiers (762 roads between Dunkirk and Menton, Le Perthus and Hendaye) which are, with the exception of those with Switzerland, internal frontiers. To control these frontiers (by the operational staff of the Central Directorate for the Repression of Illegal Immigration and Employment (DICCILEC), the customs and the Gendarmerie), each official must control about 40 kilometres. Assuming that systematic controls are maintained and that the people entering the country do so through official checkpoints (not necessarily the case for illegal immigrants), it requires the time of officials to check, including visa checks, the 230 million people crossing; according to estimates, 15 per cent of the 130 million foreigners require visas (about 20 million).[34] Unless the political regime changes, the resources in men and equipment cannot keep up with the rhetoric of systematic control. For example, the cost of the proposal to take the fingerprints of foreigners applying for resident permits has been estimated at hundreds of millions of francs. Projects to develop passports and identity cards containing microchips which would record all movements of people and provide intelligence surveillance would be even more expensive. Billions of francs would be spent for nothing – to reproduce a system which reminds us of the Berlin Wall. To give the staff of Social Security, doctors, teachers or priests a jail sentence if they do not inform on illegal immigrants would change society by creating suspicion but would not achieve the objective of ending immigration. Closing frontiers immediately generates queues lasting for many hours and even days, a phenomenon which can be observed at two land frontiers more controlled than others, the German-Polish and the American-Mexican frontiers. The United States under Nixon attempted to close their frontier with Mexico with operation 'Wetback' and, more recently, operation 'Gate-keeper', continued as operation 'Hold the line'. The result in both cases was immediate economic damage, tensions between social groups and almost zero effect on illegal immigration. The first operation was stopped after a few days, and the second was 'symbolic' since, although implemented, it simply diverted the traffic from Texas to New Mexico. The current US debate on immigration, particularly in California, demonstrates that, in the country where the ideology of the 'melting pot' reigns, a political strategy similar to that found in Europe is affecting mentalities and transforming practices.[35] The building of a 'technological wall' at Tijuana on the frontier with San Diego had a psychological impact but little practical effect. This example of reliance on technology using techniques derived from military experience – approaching the frontier like a front with strategic depth – is the product of a naive belief that machines compensate for the lack of staff.[36] The high technology system in California was jammed by the most simple counter-measures (chewing gum in machines, aluminium paper to baffle sensors).

Technologies in Europe are perhaps less military in origin, but a trend in the same direction exists. Examples are mobile checks made within the country itself, policing 'at a distance' by transferring controls into the neighbouring countries or the country of origin of migrants and European police networking.[37] Even using the most sophisticated technologies, system-atic controls at European land frontiers are impractical.[38] Although control at airports and seaports can be achieved technically, this is not the case at the land frontiers. Gérard Moreau, director of the International Migration Office, has emphasized the impossibility of conforming to parliamentary wishes to make border crossing rules more restrictive. Tightening rules reduces the number of persons eligible to enter the territory; it simply increases the number of illegal, expellable immigrants, but does not stop them getting in. The dogma of sovereignty which underpins our image of absolute control of territory by the State is associated with obsolete practices. The debates of those in parliament, of journalists, of academics ask the wrong questions until this is realized.

The contemporary paradox

Immigration depends upon millions of decisions which cannot be totally regulated by the governments without closing the frontiers. A 'double bind' is created: by deciding to control immigration completely and by aspiring to total security, we put at risk economic prosperity and political liberties associated with open societies so that the solution to immigration becomes worse than the problem. To fight against the illness, the patient has to be killed.[39] As Patrick Weil says, the problem is 'not simply the results would be unsatisfactory, but there is a contradiction between the presentation of the objectives and the implementation of policies'.[40] The structural separation between the security rhetoric and concrete measures is not due to a lack of political will to implement declarations but is explained by cynicism – it is fully understood that declarations cannot be implemented, but speeches continue to be made in the name of state security and for electoral purposes – to influence part of the electorate defined as xenophobic.

In the name of 'realism', denunciation of immigration is deemed to have a dissuasive effect – the flood of immigrants will be slowed by declaring they are not wanted. Transferring questions of immigration to the European level to avoid discussing them only amplifies feelings of unease, especially when European statements replicate national declarations and are based on the same 'myth' of control.[41] As long as the problem is viewed at European level as a problem of immigration control, and analysed almost exclusively on the basis of the notion of 'control', this prevents a wider economic, social or cultural understanding of the issues. No overall response to immigration can emerge. The main European forums on immigration are the ad hoc group on immigration set up by the Ministers of the Interior and of Justice and two dependent bodies : the *Centre d'Information et de Réflexion et d'Échange sur l'Asile* (Information and Exchange Centre for Asylum) and the *Centre d'Information et de Réflexion et d'Échange sur les Frontières et l'Immigration* (Information and Exchange Centre for Frontiers and Immigration). These have investigated all the legal methods of stopping migrants and asylum seekers by a strict interpretation of the texts and by the adding of new techniques based on dissuasion, control and surveillance. The criminalization of certain groups (immigrants, asylum seekers, diaspora, the descendants of immigrants . . . and eventually the tourists?) who, for one reason or another, cross the frontiers, seemed to some a temporary stopgap solution before controls could effectively be extended. But this latter strategy could contribute, through a security discourse, to the creation of real insecurity by a self-fulfilling prophecy that these groups are prone to criminality.[42]

A security strategy which acts as a substitute for political debate has undesirable consequences and, in this case, its basic assumptions are erroneous. Nevertheless it is increasingly important and undermines the legitimate role of European police co-operation in the field of crime prevention; police officers can thus be the victims of this strategy. Immigration control in Europe cannot continue as if it were possible to prevent those people (not as many as commonly supposed) who are ready to uproot themselves from trying their luck elsewhere. Controls can only be effective if they are based on an accurate understanding of realities, and when policy is adjusted to these realities by devices such as targeted help to some countries of the Third World, allowing legal immigration (with or without quotas) and repression of illegal employment. At the moment, illegal immigrants are surreptitiously absorbed by employers turning a blind eye. Many employers are discreet in commenting on illegal immigrants and security, in contrast to the media message that immigration is controlled by dramatic actions such as sending back scores of illegal immigrants in chartered planes. A labour force, liable to be ruthlessly exploited because they are illegal, nonetheless exists to meet the needs of some economic sectors (the building industry, public works, cafés and restaurants, the fashion and clothing industries). Government action is therefore limited to statements giving a 'feeling of security' instead of either admitting that the economy needs these people or prosecuting the employers of illegal workers.

The ability to administer systematic frontier controls belongs to the past, and probably to a mythical past. The politicization of police and security questions has not changed realities, nor has the recent turning of migration and asylum policies into matters of security. The blind alley of more and more promises of efforts of control and of the practice of targeting groups incapable of causing real political opposition can only produce negative effects. Frontiers will continue to separate societies with different levels of economic development but they no longer act as filters which designate and help to homogenize an 'inside' and an 'outside'.[43] Work, social rights, citizenship, nationality and collective identities are no longer spheres which coincide with the physical boundaries of the State. The uncertainty quality of frontier controls should not lead to a quest for total security but more reflection on possible options. A free society, it must be remembered, is one with open frontiers, open minds and plural identities. This implies both that behaviour is adaptable and that there must be acceptance of illegality at the margins. Whether European politicians accept it or not, a free society now implies tolerance of international phenomena decoupled from territory, characterized by transnational networks and the penetration of national territories.[44]

Notes

1. B. Badie and M.-C. Smouts (1993) *Le retournement du monde: sociologie de la scène internationale*, Paris: PFNSP.

2. S. Strange (1988) *States and Markets*, London: Pinter; E. Czempiel and J. Rosenau (1989) *Global Changes and Theoretical Challenges: Approaches to World Politics for the 1990s*, Lexington: Lexington Books.

3. B. Barry and R. E. Goodin (1992) *Free Movement: Ethical Issues in the Transnational Migration of People and Money*, Brighton: Harvester Wheatsheaf.

4. R. Leveau (to be published) *Projection de l'Europe à l'extérieur: espace, culture, frontière*; L. Zaki (1994) *Un monde privé de sens*, Paris: Fayard.

5. On the neologism of 'mythèmes', see J. Derrida (1992) *De la Grammatologie*, Paris: Edition Minuit. For an analysis of monopoly, see N. Elias (1964) *La civilisation des moeurs*, Paris: Calmann Lévy, and N. Elias (1993) *La dynamique de l'occident*, Paris: Presses Pocket; C. Tilly (1990) *Coercion, Capital and European States*, Oxford: Blackwell (1992, trad française, Paris: Aubier); G. Poggi (1990) *The State: Its Nature, Development and Prospects*, Cambridge: Polity. On plural identities, see B. Anderson (1983) *Imagined Communities, Reflections on the Origin and the Spread of Nationalism*, London: Verso & NLB; M. de Certeau (1980) *L'Invention du quotidien*, Paris: uge, 10/18.

6. G. Bouthoul (new edition, 1991) *Traité de polèmologie*, Paris: Payot.

7. M. Anderson (1996) *Frontiers*, Cambridge: Polity Press.

8. G. Tapinos (1992) 'L'Immigration en Europe et l'avenir des populations étrangères', *Commentaires*, Autumn, No. **59**, pp. 67–84.

9. R. Leveau (1992) 'Inquiétudes du Sud', *Esprit*, **183** (L'Europe de toutes les migrations). See also P. Perrineau and N. Mayer (1989) *Le Front National à découvert*, Paris: Presses de Sciences Po.

10. M. den Boer (1992) *The Quest for International Policing: Rhetoric and Justification in a Disorderly Debate*, Limerick: ECPR.

11. D. Bigo (1996) *Polices en réseaux: l'expérience européenne*, Paris: Presses de Science Po. M. Anderson, M. den Boer, P. Cullen, W. Gilmore, C. Raab and N. Walker (1995) *Policing the European Union*, Chapter IV, Oxford: Oxford University Press.

12. A. Nayer, M. Nys, F. Bouras and A. Maesschalk (1995) *Libre circulation, espaces européens et droits fondamentaux des citoyens*. Bruxelles: La Charte; M. King (1994) *Conceptualising Fortress Europe: a Consideration of Inclusion and Exclusion*, Madrid: ECPR.

13. Preliminary report on the Constitution for the Schengen Group. Manuscript, 1985.

14. A. Pauly Pauly (1994) *Schengen en panne*, Luxembourg: IEAP.

15. P. Masson (1996) *Rapport sur la convention d'application de l'accord de Schengen au premier Ministre*, Paris: Sénat, JORF, Documentation Française.

16. C. Harvie (1996) *Boundaries and Identities: The Walls in the Head*, Edinburgh: International Social Sciences Institute.

17. Discussion with the French Schengen Officials and the Central Directorate for the repression of illegal immigration and employment.

18. Declaration of a Member of the RPR party (Rassemblement pour la République).

19. Paul Masson, *op. cit.*

20. On imperial logic see R. Leveau (to be published) *Projection de l'Europe à l'extérieur: espace, culture, frontière*; B. Badie (1995) *La fin des territoires*, Paris: Fayard. See also Tom Nairn's chapter in this volume.

21. M. Anderson (1996) *Frontiers*, Cambridge: Polity Press.

22. D. Bigo (April 1993) *Interpénétration sécurité intérieure sécurité extérieure: enjeux américains et européens* (Report established for the FEDN; 180 pp.). M. den Boer (1994) *Rhetorics of Crime and Ethnicity in the Construction of Europe*, Madrid: ECPR; M. den Boer (1995) 'Moving between Bogus and Bona Fide', in R. Miles and D. Thränhardt (eds) *Migration and European Integration*, London: Pinter; J. Huysmans (1996) *Making Unmaking the European Disorder*, Leiden: University of Leiden; D. Bigo (1998) 'L'Europe de la sécurité intérieure: penser autrement la sécurité', in A.-M. Le Gloanec (ed.) *Entre Union et Nations; l'Etat en Europe*, Paris: Presses de Seiences Po.

23. From 25 to 65 million immigrants. For a more precise approach see figures given by M. Sehimi (1991) 'Retombée des crises maghrébines', in *Culture & Conflits*, 2: *Menaces du Sud: images et réalités*. See also the last works published by the INSEE (French National Institute of Economic and Statistical Information).

24. A reading of the Convention and the obligation of declaring foreigners shows that there is no laissez-faire policy, no laxness, rather the contrary. This does not stop Jean-Marie Le Pen declaring that 'Schengen opens the doors to drugs and insecurity as well as to immigrants and refugees from all over the world'.

25. Senate (28.06.1991) Debate on ratification.

26. Inquiry by the Parliamentary Committee. Jean-Pierre Philibert (president), Suzanne Sauvaigo (rapporteur) (1996), *Immigration clandestine et séjour irrégulier d'étrangers en France*, p. 43, part I.

27. On the link between physical and symbolic frontiers see Anthony Cohen's chapter in this volume.

28. D. Bigo (1995) 'Troubler et inquiéter: les discours du désordre international', in *Culture & Conflits*, No. **19/20**, pp. 19–20.

29. Documents from ECRE and *Amnesty International* (June 1991).

30. As well as ECRE, we can mention GTEEAR: Groupe de Travail des Eglises Européennes sur l'Asile et les Réfugiés (Working Party of the European Churches on Asylum and Refugees).

31. *Fortress Europe*: Newsletter from N. Busch. See also B. Leuthardt (1995) *Festung Europa*, Zürich.

32. M. den Boer (1995) 'Moving between Bogus and Bona Fide' in Miles and Thränhardt (eds) *Migration*.

33. F. Ewald (1996) *Histoire de l'Etat providence*, Paris: Le livre de poche, p. 17.

34. The figures are taken from various documents which were given at the time of discussions with members of DICCILEC (Central Directorate for the Repression of Illegal Immigration and Employment), the DLPAJ (Direction des Libertés Publiques et de l'Administration Judiciaire au Ministère de l'Intérieur: Directorate for Public Freedom and Judicial Administration at the Ministry of the Interior), the OFPRA (Office de Protection des Réfugiés et des Apatrides: Protection Office for Refugees and Stateless Persons), the DCPJ, the officials of the Schengen system. See also Michel Foucher's chapter in this volume.

35. Now we say 'salad bowl' and no longer 'melting pot'. We blame multiculturalism for all 'evils'.
36. J.-P. Hanon (October 1996) 'L'armée veille à El Paso', *Le Monde diplomatique*.
37. Bigo, *Polices en réseaux, op. cit.*, Conclusion.
38. In total, according to the estimations of the French co-ordinator for free movement, there would be 1.2 billion crossings per annum at the Schengen frontiers, in other words 3.287 million crossings per day.
39. P. Watzlawik (1988) *Comment réussir à échouer, trouver l'ultrasolution*, Paris: Seuil.
40. Patrick Weil has demonstrated in a very convincing way that the migration breaker is based on imagination, on fears and not on an objective appreciation of figures. Cf. (1992) *Logiques d'Etats et immigrations*, Paris: Kimé. See also the note of the *Foundation Saint Simon* (April 1996) in *Esprit*, entitled 'Pour une nouvelle politique d'immigration'.
41. D. Bigo (to be published) *Fortress Europe or Policing at a Distance?*, University of Leiden.
42. D. Bigo (1993) *Du discours sur la menace et de ses ambiguïtés*, Paris: Cahiers de l'IHESI, No. 14.
43. Anderson, *Frontiers*.
44. B. Badie and M.-C. Smouts (1996) *L'international sans territoire*, Paris: L'Harmattan; D. Bigo (October 1996) 'L'illusoire maîtrise des frontières', *Le Monde diplomatique*.

Chapter 11

The New Frontiers of European Policing[*]

NEIL WALKER

Introduction

The recent growth of a European transnational policing capacity is undoubtedly connected with the broader dynamic of political development within and beyond the European Union. It is widely acknowledged that the acceleration of the integration project in the 1980s and, in particular, the commitment in the 1992 Single Market Programme to the removal of national frontier controls on persons and goods created a political climate favourable to EU initiatives in policing and other security matters. It is less well understood, however, that the relationship between broader trends in European politics and the emergence of a policing capacity is multi-faceted and marked by internal tensions. In turn, this has brought about a situation in which the contemporary record of European policing is one of uneven development and the future remains shrouded in doubt. In particular, uncertainty surrounds the relationship between the law and formal policy of the EU in policing matters and its actual practice, the prospects for police accountability and the ultimate location of authority for policing in a post-sovereign Europe.

We will address the causes and manifestations of this uncertainty in due course, but first we set the scene by signposting the main developments that have taken place to date. Considerations of space dictate that only the sketchiest of overviews is provided, but the interested reader can consult any number of sources to establish a more detailed picture.[1]

From Trevi to EUROPOL

The formal recognition of the relevance of a policing dimension within the European Community came rather late in the day and, as with the project of European integration more generally, followed a rather uneven course of development. The Trevi organization provided the first major initiative in the policing field in 1975, a full eighteen years after the signing of the Treaty of Rome. Trevi served as an intergovernmental forum for Member States to develop common measures with regard to counter-terrorism and, later, with regard also to drugs, organized crime, police training and technology and a range of other matters. The intensification of Trevi's activities coincided with and was given additional impetus by the development of the European Commission's Single Market Programme after 1985. Whereas previously Trevi had been isolated from the mainstream of Community policy making, now its security concerns were integral to the doctrine of compensatory measures through which the Co-ordinator's Group on the Free Movement of Persons sought to respond to the 'security deficit' implied by the opening of internal borders.[2]

The Schengen system, which provided a second important step towards a European law enforcement capacity, is even more closely associated with the Single Market initiative. The initial Schengen Agreement of 1985 was the vehicle of an inner core of five member states who wished to take the initiative in the abolition of border controls and so accelerate progress towards the completion of the internal market. A more detailed Implementation Agreement of 1990 established a number of associated law enforcement measures including the Schengen Information System and police co-operation in matters such as 'hot pursuit', cross-border observation and controlled delivery of illegal goods. After a succession of false starts, the Schengen system eventually became operational in March 1995, although a zone entirely free of permanent border controls has yet to be achieved amongst the (now) seven fully participating members. Despite these problems, Schengen's sphere of influence grows ever wider. All except the United Kingdom and Ireland of the fifteen Member States of the EU are now members or intend to join, while the two non-EU members of the Nordic Passport Union, Norway and Iceland, are to be granted associate status.[3]

While Trevi and Schengen represent noteworthy advances in the development of a European law enforcement capacity, the European Police Office (Europol) constitutes the most ambitious plan yet conceived in this field. Its foundations are to be found in the Maastricht Treaty on European Union of 1992, which included policing within a new Third Pillar of EU competence over Justice and Home Affairs. Whereas Trevi, to which the Third Pillar is the institutional successor, was a purely intergovernmental arrangement located at the margins of the Community, the Third Pillar is (to continue the architectural metaphor) built into the masonry of the new European Union. The major European institutions – Commission, Council, Parliament and Court of Justice – may play a less central role than they do in the traditional First Pillar, where the Union has full legislative, executive and judicial authority. Nevertheless, the Council and Commission in particular continue to exert a significant influence over the decision-making process alongside the individual Member States.[4]

Under the Third Pillar, an elaborate structure has been put in place spanning five organizational levels. Beneath the two familiar Community executive organs – the Council and Coreper – lies the K4 Committee, which comprises permanent representatives of the justice and interior ministries of the Member States. In turn, the K4 Committee has generated a number of sub-groups which constitute the fourth and fifth levels of the hierarchy. Police and customs co-operation provides the remit of one of the three steering groups, the others dealing with immigration and asylum and with judicial co-operation in criminal and civil matters respectively. Like the other two, the police and customs co-operation steering group has generated a number of specialist working groups, including groups on terrorism, drugs and organized crime, co-operation of a scientific and technical nature and the preparations for Europol itself.

A number of initiatives have emerged from this elaborate structure, but the development of Europol has been the focal point. It has also provided an instructive example of the uneven course of progress under the Third Pillar generally. The foundation provision in the Maastricht Treaty envisaged Europol as a system of information exchange for the purposes of preventing and combating terrorism, drug trafficking and other serious crimes within the EU and as a means of providing co-operation in aid of criminal investigations and analyses more generally. It was clear from the outset, however, that notwithstanding Chancellor Kohl's strong support for the prompt establishment of Europol as a full-scale criminal intelligence agency, progress would be achieved only in small increments.

Despite an early post-Maastricht commitment to provide Europol with the full legal authority of a Third Pillar convention,[5] in the short term the main focus of institution-building was upon the more modest and prefigurative Europol Drugs Unit (EDU). But even here the timetable slipped. The EDU agreement was originally intended to be completed by the end of 1992, and after it was eventually signed in June 1993 it took another five months until the location of the headquarters of the new unit, the very issue at the heart of the initial delay, was resolved in favour of The Hague. Further, the initiative originally assumed the highly informal character of a Ministerial Agreement, and it was not until 1995 that it was accorded formal recognition under the Third Pillar as a joint action. By that time, the remit of the fledgling organization had been extended beyond drug-trafficking and associated criminal organizations and money laundering activities to cover nuclear crime, illegal immigration networks and international vehicle crime. However, even this expansion is the legacy of thwarted ambition, as it took place at the initiative of the German presidency of the European Council which had been embarrassed by its failure to achieve a target of agreement on the full Europol Convention by the end of 1994.

The Europol Convention was not finally signed until July 1995, following protracted debate over the mandate and powers of a mature Europol, the adequacy of its data protection systems and its framework of legal and political accountability. Even then, the process of ratification and implementation could not be completed until the outstanding question of the role of the European Court of Justice in interpreting and enforcing the Convention was resolved. The United Kingdom had remained implacably opposed to recognizing any role for the Court throughout the negotiation of the Convention, and when a solution was eventually found in a protocol of July 1996, it effectively took the form of an agreement to differ. The United Kingdom would not itself recognize the jurisdiction of the Court, but acknowledged the right of the other fourteen member states to accept its jurisdiction over a limited range of questions.

There remains, however, a number of obstacles to full implementation of the Convention. Although Europol rests on a more secure constitutional foundation than the EDU, it shares with the latter a gradualist approach to the extension of its powers, its original remit largely confined to those matters already covered by the EDU.[6] More generally, the Convention takes the form of a framework document, and a series of secondary rules and regulations require to be approved before the system can be fully operational. Finally, technical difficulties and continuing uncertainties over its precise role in linking national units have led to predictions that the Europol computer system – the fulcrum of the new organization – will not be viable until 1999 at the earliest.[7]

Undercurrents of police co-operation

How do we account for the unsteady progress towards police co-operation in the EU? Building upon an earlier analytical framework, I intend to shed light on the pattern of development by reference to a number of factors.[8] In the first place, there is a series of public political discourses in terms of which arguments for and against police co-operation have been framed. In the second place, there is a set of underlying influences which are unlikely to be articulated in political debate but which nonetheless remain influential.

Public political discourses

The most general public conversation and debate affecting the development of European police co-operation is the discourse of European integration itself. According to the affirmative model of European integration, the EU is seen as institution which transcends the state and which has a political authority and range of competences to rival that of the state. And to the extent that the project of European integration is conceived of favourably, the invocation of this broader project will also support greater integration in particular areas, such as policing and Justice and Home Affairs (JHA) matters more generally.

Thus it is no mere coincidence that the Third Pillar and Europol initiatives took place at the high tide of European institutional self-confidence in the integration process, following the Commission's audacious success in marketing the 1992 project. Indeed, in July 1991, the Dutch draft of the EU Treaty – which was eventually defeated – took advantage of this receptive political climate by proposing that the planned JHA chapter be fully integrated within the existing Community framework. Chancellor Kohl's announcement of a proposal for a European Police Office (Europol) in 1991 was similarly in tune with the political mood, typifying the ambitions of the increasingly dominant EU Member States at a point when a significant deepening of the integration project appeared politically viable.[9] It was intended, and partly succeeded, as a potent symbol of European Union precisely because it was so audacious, promising to transfer authority in an area which was one of the most traditional and closely guarded preserves of the state.

Of course, the discourse on European integration often favours the opposite conclusion. The conservative emphasis upon state sovereignty represents a major obstacle in the path of any major additional policing competence at EU level. In part, this is a matter of the general health of the integrationist project. Just as its overall robustness assists particular areas of co-operation, so any general loss of momentum can undermine progress in particular spheres. For instance, since the difficulties which attended the ratification of the Treaty on European Union (TEU) announced a resurgence of nationalist sentiment against a strong EU, we have witnessed in the last few years a trimming of integrationist ambitions in all areas. This can be detected on a broad canvas, as in the markedly less-ambitious terms of debate for the present Intergovernmental Conference compared to its predecessor prior to the TEU.[10] It can also be discerned in detailed cameos. One graphic illustration is the blocking policy pursued by the British government in response to the ban imposed by the European Commission on British beef products in the wake of the BSE scare in the early months of 1996. While it lasted, this strategic stand-off, which provided an occasion for a more general restatement and reinforcement of opposing positions in the integrationist debate, led the United Kingdom to suspend negotiations on a wide range of EU business, including the unresolved Europol Convention.[11]

Aside from the general current of Euroscepticism, from within the state-centred perspective there is a more specific concern with the appropriateness of supranational developments in the JHA area. As suggested, law, order and criminal justice have a close traditional association with domestic politics.

The reasons for this are complex, but concern the circumstances of the emergence of the modern state as a way of guaranteeing security from within and beyond territorial borders, and as a means of providing its inhabitants with a political identity and allegiance in the form of citizenship, together with a corresponding set of rights, benefits and obligations.[12] Policing, particularly in areas such as drug-trafficking and terrorism where international co-operation is at such a premium, is heavily involved in protecting the 'specific order' of the state as well as the rights and benefits of the citizen which flow from the maintenance of 'general order'.[13] Other elements of the JHA agenda are similarly implicated. Immigration and asylum policies concern the identity of those permitted to travel or reside within the territorial unit of the state. For its part, the politics of border controls are located precisely at the point of intersection between issues of security and identity. Overall, the symbolic link between the political sphere and JHA matters is well-grounded in the history of the modern state and continues to provide a significant check on integrationist ambitions in this area.

While pro-integration arguments tend to favour police co-operation, and anti-integration arguments tend to act as a brake, this does not always follow. It is instructive, for example, that states such as Denmark and, in particular, the United Kingdom, which have been prominent members of the Eurosceptic camp in recent years, have remained broadly supportive of enhanced police co-operation. As we shall see, this is in part due to other arguments in favour of co-operation. There is also, however, a sense in which a cautiously affirmative position in respect of police co-operation aids the development of a plausible alternative to a strongly federalist position within the broader integration debate. Support for an issue as close to national security concerns as police co-operation helps to anchor a policy of pragmatic intergovernmentalism. To its exponents, the attractiveness of such a policy, and of the image associated with it, is that it eschews the uncritical generality of a broadly pro-federalist stance and, equally, distances itself from the similarly indiscriminate stance of the Eurocynic – suspicious of all forms of co-operation and committed to an insular nationalism. Instead, pragmatic intergovernmentalism claims that every issue should be looked at on its merits, and that where national interest so dictates, co-operation should not be dogmatically rejected.[14] Of course, given its inherent opposition to a federalist end-state, such a position can only operate within strict limits, and is apt to reject any general concession to the authority of European institutions. This balance is clearly represented in the British government's position paper on the 1996 IGC, which is both enthusiastic about more efficient and more intense Third Pillar co-operation and implacably opposed to a greater role for Community institutions.[15]

A second public discourse closely associated with the development of European policing is the discourse of *functional spillover*. Since its earliest development, a key argument associated with the extension of the European project into new domains is that for intervention in one sector to be optimized requires adjustments in other related policy sectors.[16] In particular, the development of an efficient and effective common market – the founding mandate of the European Community – required spillover policies as diverse as anti-discrimination measures to ensure a level playing-field within the labour market and common welfare and environmental measures to ensure some equivalence of 'externalities' across enterprises located in different states. The argument about the need for security measures to compensate for the opening of internal frontiers is one more instance of the functionalist theme, and serves to demonstrate how the logic of functionalism is generally favourable to enhanced police co-operation.

There are other, more subtle ways in which functionalist patterns of thinking insinuate themselves into the debate about police co-operation. To begin with, functionalist arguments can be applied at the micro-level of Community activities as well as providing one of its macro-political narratives. Arguments for the development of an administrative policing capacity and quasi-criminal sanctions within the EU in areas such as the enforcement of competition law and the prevention of fraud against the EU are of this type.[17]

Second, arguments in favour of policing, including international policing, may be persuasively couched in functionalist terms, even where such arguments are not specifically tied to the policy framework of the EU. As Cohen observes, given their flexible capacity to provide an authoritative solution and their easy accessibility, the police tend to exercise a 'stand-in' authority[18] within all polities, plugging the gap wherever the normal authoritative solution or practice has failed. Or in functionalist language, there is inevitable spillover into the policing domain from a wide range of other policy sectors or social activities. Whether the argument is that the growth of business enterprise across Western and, increasingly, Eastern Europe leads to a corresponding increase in crime connections, or that the development of international leisure has produced the international football hooligan and the international paedophile ring, or that the information revolution has led to increased use of computers in the commission of international crimes, there is a persuasive case to be made for policing measures to compensate for dysfunctions in other sectors. That case is all the more persuasive in the European field in that it resonates so closely with the specialist functional logic which underpins so much of the debate about the expansion of the EU.

For present purposes, the truth-value of these arguments is of less significance than their political efficacy. Indeed, this point is underlined by the fact that with regard to the broad functionalist argument about the compensatory security requirements of the opening of internal borders, neither the premise nor the conclusion bear close critical scrutiny. In 1996, the Community is yet to pass measures to eliminate border controls, and even the draft Directive on this matter, proposed by the Commission in 1995, makes significant concessions to public security arguments.[19] In any case, the link between the removal or relaxation of such controls and growth in international crime remains largely unproven.[20]

Nevertheless, functionalist reasoning of this sort continues to be influential because it appeals to a technocratic vision of public policy as the performance of a range of necessary but neutral tasks. Internally, this is in harmony with the working culture of many state (and suprastate) functionaries, including the police with their self-image of a technologically enhanced,[21] pragmatic professionalism.[22] Externally, the success of functionalism rests on its capacity to present the integration process as other than, and as 'ideologically transcendent over the normal debates on the left–right spectrum'.[23] The language of functional spillover, therefore, is much more than a (largely discredited) theoretical hypothesis purporting to *account for* change. It is also a powerful, if often understated, ideological currency which taps deep reserves of pro-European sentiment and which can be effectively harnessed in aid of integration.

The relationship between the discourses of European integration and functional spillover is a complicated one. In the broadest sense, they represent the tension within the foundations of the European project between the wide political vision of the new Europe and the narrow economic calculation of a common market. One promotes integration as a new kind of meta-political project, while the other trades on its denial of grand politics. Furthermore, the functional argument in favour of integration is often forced to proceed against the grain of political opinion in the wider debate.

On the other hand, in appropriate circumstances these discourses can operate in a mutually supportive fashion, and can do so in a manner which either favours or opposes closer integration. As shown by the way in which the European Commission, and in particular its then president Jacques Delors, utilized the technocratic vehicle of the 1992 Programme to develop a new expansionist perspective,[24] the logic of functional spillover can be presented as contributing to the uniqueness of the European political project, rather than as a denial of its political character. Conversely, functionalist arguments may less frequently be invoked in aid of an anti-integrationist perspective. For example, the kind of pragmatic intergovernmentalism favoured by the United Kingdom government in recent years as an alternative to a more ambitious federalism is often backed up by a modest functionalism.

A third public political discourse which, as in the case of functional spillover, tends to be supportive of enhanced JHA integration, is the discourse of internal security.[25] Internal security discourse is best understood as part of a wider project which focuses on the security of Europe as crucial to its well-being as a political entity. Two concerns underpin this project. First, there is a search for a cultural, or affective identity. This has undergone a strong revival since the late 1980s. The significant resurgence of the politics of national identity in fragmented Eastern Europe, the ambitious programme of territorial expansion of the EU and the searching questions asked of the legitimacy of an incremental drift towards a more federal Europe during the ratification of the TEU, all served to focus attention on the question of the cohesiveness of European civil society.[26]

If, as is often contended in response to the narrow vision of the functionalists, the cultural sphere lags behind the political and economic spheres in the integration process, it is hardly surprising. Europe lacks the vivid, readily accessible and well-established mythology and symbolism from which a common identity is typically constructed at national level. Instead, the development of a European identity is a precarious exercise. It involves a piecing together of various 'conceptual fragments'[27] such as Europe as a territory, as a model for the balance of power, as the cradle of liberty, as civilization and as Christendom. The absence of a central definition within this vision, together with its tone of cultural imperialism, means that the new cultural politics accentuates a theme present in all identity politics, namely the idea of exclusion and otherness.[28] It is easier to say what Europe is not, rather than what it is. 'Europe' is not Eastern Europe; it is not Islam; it does not extend to its former colonies; it is certainly not continuous with its own fragmented and frequently warring past.[29]

There is also a key economic strand within the development of Europe as a security community. The stable prosperity of the EU stands in stark contrast with the insecurity and poverty of the polities and economies to the east and south. In particular, the end of the Cold War means that the fruits of the West are no longer forbidden. In these circumstances, exclusion and resistance to migratory pressures has come to be seen as prerequisite to the maintenance of Western Europe's economic advantage.[30]

According to this view, Western Europe must be protected against cultural and economic threats from alien forces within or beyond its borders. There is a growing tendency for these issues to be merged and treated as a single problem of security. The political advantages of translating complex problems into the language of security are well stated by Waever:

> Security is a practice, a specific way of framing an issue. Security discourse is characterised by dramatising an issue as having absolute priority. Something is presented as an existential threat: if we do not tackle this, everything else will

be irrelevant (because we will not be here, or not be free to deal with future challenges in our own way). And by labelling this a security issue, the actor has claimed the right to deal with it by extraordinary means, to break the normal political rules of the game (for example, in the form of secrecy, levying taxes or conscripts, limitations on otherwise inviolable rights). 'Security' is thus a self-referential practice, not a question of measuring the seriousness of various threats and deciding whether they 'really' are dangerous to some object. . . . It is self-referential because it is in this practice that the issue becomes a security issue.[31]

Arguing along similar lines, Bigo has claimed that the elaboration of an 'internal security ideology' has allowed a number of concerns arising out of the basic cultural and economic anxieties, from terrorism to immigration, and from organized crime to asylum to be located along a single 'security continuum'[32] and treated to a one-dimensional response. On this view, it is nothing less crucial than the security of the Western European way of life that is the first and foremost priority across a range of JHA policies, from reinforcing the hard outer shell of the EU's external frontiers to conducting systematic internal checks and developing an autonomous policing capacity.

Of course, there exist other, less apocalyptic ways of presenting and practising identity politics within the EU. The emphasis upon security and exclusion exists alongside a strong tradition of liberal toleration and abhorrence of racist attitudes and politics. Indeed, the recent Joint Action to combat racism and xenophobia, agreed in early 1996 after protracted controversy, indicates that policies consistent with this tradition can emerge from the Third Pillar – the very engine room of internal security. But if this suggests the co-existence of the two perspectives, it is not a co-existence of mutual engagement and compromise but rather one of incommensurable paradigms of thought. In focusing on the immediacy of the threat posed and pandering to the 'respectable fears'[33] of the European middle classes, the 'self-referential' internal security discourse closes off other more complex and considered ways of perceiving the various objects of its concern. In particular, it avoids the need to confront head-on unpalatable questions about growing inequality between the world's regions and also succeeds in hiving off the problems of combating European racism and of protecting fundamental personal rights to separate policy compartments, ostensibly unconnected with the mobilization of public resources against external dangers.

Just as there are tensions between the discourse of European integration and that of functional spillover, so there may also be tensions between the discourse of internal security and the other two. The discourse of internal security involves a more negative and short-term approach than the assertive and visionary discourse of European integration and wears its political commitments more openly than the discourse of functional spillover. Also, as with the discourse of functional spillover, the extent to which the discourse of internal security can sustain support for a strong version of integration may be limited. As Waever acutely observes,[34] the meaning and usage of the notion of security is historically bound up with the idea of the nation-state. The fact that Europe is increasingly becoming a referent object of security, not only in an internal sense but also in an external sense – as an actor on the world stage – suggests that we are beginning to invest the idea of Europe with many of the characteristics and attributes previously attributed to the state. However, this is uncharted territory and the next stage of the journey is not clear. While it is now generally conceded that the nation-state is no longer the exclusive focus of security concerns, the idea of security remains central to its identity and integrity. If the idea of European security begins to assume such proportions that it is perceived or portrayed no longer as supportive of nation-state security but rather as a threat to it, there remains the possibility of a backlash and the retrenchment of the nationalist perspective.

Underlying influences

In addition to the public political discourses, two sets of underlying influences are prominent in supporting the development of a significant European policing capacity. In the first place, there is the development of an elaborate structure of transnational bureaucratic interests around policing and JHA matters generally.[35] The background to this is the long history of practical co-operation between policy professionals and administrators which predated a resort to formal political structures. This legacy remains strong, for although the Third Pillar is innovative in many respects, there is a significant degree of continuity of tasks and personnel from its predecessor organization, Trevi, while a whole range of other traditional co-operative practices have been uninterrupted by the Maastricht watershed.[36]

The strengthening of transnational professional and bureaucratic bonds has been reinforced by two aspects of JHA politics. First, the success of spillover arguments and the robustness of the internal security ideology have provided a supportive context for the incremental linkage of a wide range of issues within and beyond policing, so contributing to the intensification and expansion of Third Pillar activity. In the second place, continuing uncertainty over the future of European security policy has left a number of other security and criminal justice agencies, including Schengen, Interpol and the Council of Europe, jockeying for a strategically advantageous position to accommodate the security needs of tomorrow.[37] The pattern of activity within such an uncertain market need not lead to effective competition and rationalization, but may instead produce ever more narrow niches of expertise and the mounting of large, overlapping agendas by rival organizations seeking to anticipate all possible future scenarios.

Domestic political imperatives also provide underlying support for the development of a European policing capacity. This proposition may appear to sit uneasily with the often publicly adversarial climate of debate between national and European political elites. Beneath this appearance of discord, however, there is frequently a concurrence of interests. If we conceive of the national level as compromising a number of sectoral policy communities[38] rather than a single voice, then particular departments of government may develop significant common ground and good working relationships with their European domestic and supranational counterparts, notwithstanding any positions adopted or postures struck by the same domestic governments when operating in collective mode.

More specifically, the addition of a European dimension may be welcomed by the domestic representative of a particular policy community as corroborating the continuing viability and efficacy of the domestic policy contribution.[39] Caught in a repetitive cycle of policy options, the political masters of domestic criminal justice agencies across the EU are often compromised by their increasingly apparent failure to sustain the myth of 'sovereign crime control'[40] within national boundaries, and to struggle to find adequate justifications for new initiatives. In contrast, developments in the European security field, including crime trends and patterns of migration, may be invoked as factors which lie beyond the control and responsibility of domestic agencies, and which provide a clean slate on which to address fresh challenges and mobilize renewed support. For example, the separate European agenda on crimes such as terrorism, drug-trafficking and money-laundering may be, and has been, effectively invoked to justify the allocation of domestic resources towards that European effort or to domestic activities in tandem with it. Likewise, the development of the external frontier as a key reference point for European security policy creates a new set of priorities and rationales for national immigration and asylum policies and for internal security checks.

The uncertain future of European policing

The emergence of a significant European policing capacity, therefore, has been determined by a wide range of factors. Functional arguments, internal security concerns and the interests of transnational bureaucratic networks and domestic policy communities may drive the European policing and internal affairs agenda so far and so fast, but its momentum is apt to be interrupted on two scores. In the first place, the support offered by functional arguments or internal security arguments is not entirely unequivocal. Both can be harnessed to anti-integrationist agendas and, even where supportive, endorse visions of a European future which may appear to many of their audience uninspired or unattractive. In the second place, national sovereignty concerns – the ideological flipside of the strong integration argument – are available, and in recent years regularly resorted to as a powerful brake whenever any state passenger becomes apprehensive at the rate or direction of progress.

In this final section, it will be argued that the complex interaction of forces which has produced an uneven contemporary record also suggests an uncertain future for European policing. The argument will proceed through consideration of three related themes.

The gap between legal form and operational practice

For many years, sociologists of law[41] and sociologists of organizations[42] have cautioned us to view formal legal and organizational rules in a sceptical light, and to expect to discover a significant disparity between the rule book and the 'rules in action'. This may be due to a number of factors. Written rules serve symbolic functions, announcing public commitments which may not be honoured in practice. Rules may also be used instrumentally, as a resource to achieve the independent ends of the rule-user rather than as a normative guide. Rules may also be frustrated or overreached by those to whom they nominally apply, because they have both the will and the wherewithal to avoid them or outstrip them. In the context of European police co-operation, such is the complex interaction between the different factors bearing upon development that all these possibilities come into play.

To begin with, there is a strong pressure towards the symbolic affirmation of consensus in the international policing domain, leading to the premature conclusion of agreements. The Schengen Implementation Agreement, the Ministerial Agreement founding the EDU and the Europol Convention itself are all examples of agreements which were signed before important matters of principle had been resolved and where a considerable delay could be anticipated prior to implementation.

In part, this represents a compromise between pro-integrationist and anti-integrationist tendencies. On the one hand, there is a formal record of progress to add to the integrationist narrative. On the other hand, as with the unresolved status of the European Court of Justice under the original Europol Convention, the terms of the text guarantee that there can be no practical concession to the integrationist position pending a decision on the matter of principle.

Premature agreement can also serve the pragmatic interests of the functionalists, the crisis-management tendencies of the exponents of the internal
security approach and the practical agendas of transnational bureaucracies
and national policy communities. As the busy preparatory schedules undertaken by the various new European police organizations prior to their official
commencement date indicate, the public affirmation of an international
commitment to action, however premature, provides a sense of permission
for those committed to enhanced co-operation to proceed in anticipation of a
final resolution of outstanding disputes.

The gap between the rule book and practice is perhaps even more starkly
evident in the tendency of practice to outstrip formal powers even where no
framework document is available to provide a legitimating gloss. Again, this
tendency is attributable to the pragmatic, practice-driven agendas which
functionalist and security arguments encourage and which powerfully
entrenched national and transnational interests are well placed to pursue. In
turn, the ascendancy of the operational imperative can lead to a relentless
incrementalism whereby important thresholds in co-operation may be
passed without public debate or formal acknowledgement.

For example, although the debate over the Europol Convention made clear
that Europol should remain an agency without its own executive powers
(arrest, search, questioning, etc.) or operational competence, but should
instead provide a service function in matters of information exchange and
intelligence analysis to national operational units,[43] the balance of power
and responsibility between national and European policing levels appears to
be undergoing a subtle but important transformation. Jürgen Storbeck, the
co-ordinator of the EDU, has argued that there is a fine and flexible line
between the co-ordination of intelligence between national units, for which
Europol will be able to take full responsibility once the Convention is fully
implemented and personal data can be stored centrally, and the conduct by
Europol of its own investigations.[44] The crux of the argument is that in its
intelligence co-ordination role Europol will acquire a unique overview of
certain transnational crime patterns and incidents, and thus will be better
placed than any other policing units to respond operationally. And in the
absence of its own executive powers, Storbeck argues that through the
establishment of ad hoc international task forces, Europol can direct (and,
indeed, on a modest scale, has already begun to direct) investigations by
nationally competent officers in their several states towards an overall
resolution of a particular case. The significant slippage from a service role to
a directive role is justified on operational grounds as the inexorable result of
Europol's developing resource and intelligence base, and is motivated by a
mixture of empire-building and pragmatic professionalism.

The wider point to be drawn from this analysis of the gap between theory and practice is that it presents a very uncertain picture of the progress of European policing to a number of interested constituencies. That includes those concerned as commentators to understand and critically appraise developments in European policing, those concerned as legislators and regulators to sustain a level of control over the developing institutional framework, and those concerned as members of the public that the formal record of what is done in their name should remain closely attuned to operational practice. To all these groups, the relationship between form and practice remains attenuated and opaque, and given the present constellation of forces driving developments in European policing, it is difficult to see how this will change in the foreseeable future.

The uncertain accountability of European policing

The gap between the official form and actual practice of European policing raises questions about the transparency of working arrangements and the relevance of regulatory oversight which clearly cast doubt on the efficacy of the systems in place for holding European policing to account. These doubts and uncertainties can be placed in sharper focus, however, if we examine in a more detailed fashion the relationship between each of the public political discourses and underlying influences and the development of structures of public accountability.

Neither the discourse of functional spillover, with its privileging of techno-cratic expertise and its denial of grand politics, nor the discourse of internal security, with its emphasis upon the uncompromising pursuit of a narrow objective, is particularly receptive to the idea of an elaborate framework of external accountability, with detailed overview of tasks and consultation of a wide range of opinions. Further, the transnational bureaucracies and specialist national policy communities whose interests are presently favoured are disinclined to upset the prevailing balance of power by exposing the decision-making processes within European policing to external scrutiny and influence.

If this helps to explain why Europol, like its predecessors and competitors, allows only a modest role to those institutions best placed to exercise an external overview of its functions – in this case the European Parliament and the Court of Justice[45] – it does not account for the lack of resistance to this prospect by the strong anti-integration forces within the wider ambit of European politics. After all, surely the Eurosceptics have particular reason to be wary of the empire-building of powerful and relatively unaccountable Third Pillar institutions. The answer to this conundrum lies in the fact that from the perspective of national sovereignty, the proper lines of account-ability are through national channels, which means domestic parliaments and domestic courts. On this view, the strengthening of any European institutions, even those which are designed to reduce the deficit in account-ability and democracy, may be perceived in narrow zero-sum terms as contributing to the erosion of national self-determination, and for that reason resisted.[46] There is, in other words, a refusal – or at least a reluctance to contemplate that the locus of power has already shifted so much to the European centre that the accountability system must adapt accordingly, even if that entails empowering particular European institutions. Thus the narrow pragmatism and the bureaucratic conservatism which provides much of the impetus for European police co-operation conspires with the profound scepticism of anti-integrationist forces to provide a scenario in which the development of strong mechanisms of accountability remains a highly dubious prospect.

Policing the post-sovereign Europe

We have seen how the issue of police co-operation raises profound questions about the distribution of sovereignty within the new European Union. The speed and scope of development of the new European policing capacity and the nature of its accountability arrangements all are influenced by and in turn influence the debate between pro-integration and anti-integration forces about the proper allocation of sovereign authority.

Gradually, however, it would appear that our traditional ways of framing questions about sovereignty are becoming inadequate to the complexity of the development of the European Union within the wider international community. Recent initiatives such as the European Economic Area (EEA), the Europe agreements with candidate states, the Social Charter and EMU opt-outs of the Maastricht Treaty, and the general discussion of flexible integration within the 1996 IGC have reflected the development of a less-rigid attitude to questions of institutional architecture and the relationship between Union and non-Union institutions.[47] The reasons for this are many and varied, but the most important is that in the current finely balanced political debate over the future of Europe, flexible integration as a broad strategy appeals to both pro-integration and anti-integration forces. For those in favour of greater integration, flexibility avoids the dangers and frustration of institutional stagnation, with the overall European project proceeding at the pace of the most reluctant Member State. For those opposed to greater integration, flexibility promises a more loosely structured set of arrangements. On the latter view, if states are allowed to co-operate on an à la carte basis, with different coalitions around different concerns, not only will this arrangement be more respectful of divergent national interests, but it will also begin to dilute the significance of the Union as a distinctive and cohesive entity with particular interests and a common set of institutions.

Whichever view is more plausible, a situation is clearly emerging where the debate over European sovereignty can no longer be perceived in the traditional vein as the relationship between two fixed entities – the EU and the Member States. The EU increasingly means different things in different sectors, while the Member States, together with a number of non-member states, now find themselves in variable relationships with the institutions of the EU. In order to understand this changing mosaic, we need to forge more subtle and flexible concepts of political authority.

The move towards flexible integration in a post-sovereign Europe applies as much, if not more, to policing and JHA matters as it does to any other area of Union activity. This is demonstrated by the opt-out on the jurisdiction of the Court of Justice negotiated by the UK under the Europol Convention. It is also indicated by the general willingness of the IGC negotiators, recorded in the Florence Summit of June 1996, to accept the idea of flexible integration more readily in the context of the Second and Third Pillars than in the traditional sphere of the First Pillar, where the perceived need to preserve the essentials of the single internal market reduces the scope for differentiation.

The attractions of a more flexible and diversified approach in policing and Third Pillar matters is also apparent from the spate of security agreements and liaisons recently negotiated between the EU and other states.[48] These include the Barcelona Declaration of 1995, which paved the way for collaboration with the Maghreb countries of North Africa to control migration and curb illegal immigration at the Southern external frontier of the EU; the EU–US Summit of December 1995, which declared common cause and anticipated a common programme of action between the two entities in the fight against international crime; the meeting of the new P8 group in Ottawa in the same month, where the eight most industrialized nations, including four members of the EU, determined to take co-operative measures against terrorism; the inter-regional trade co-operation agreement between the EU and the countries of Latin America of 1995, which included a commitment to co-operative measures against drug-trafficking; and the decision of the 1994 Essen European Council to commit funds from the PHARE programme in aid of democracy in Central and Eastern Europe to the treatment of Third Pillar matters, a commitment which led to the Langdon report to the JHA Council in 1995 recommending more intense police co-operation between the two groups of states and the recruitment of central European 'buffer states' to the project of reducing clandestine immigration from the east.

Again, these various developments are understandable in terms of the public political discourses and underlying influences bearing upon European police co-operation. We have already seen how both sides of the master debate on European integration might favour more flexible developments, and both the final form of the Europol Convention and the emerging mood for a more flexible approach to JHA matters in the 1996–7 IGC are cases in point. Functional arguments may call for interventions beyond the territorial sphere of the EU in order to protect developments within the EU. Those who subscribe to the ideology of internal security are faced with the paradox that they may have to enlist the support of external forces in order to strengthen the security of the EU, and initiatives such as the Barcelona Declaration and the Langdon report suggest that faced with this choice the argument for external collaboration often prevails. For transnational bureaucratic interests, lack of flexibility in an uncertain market is a drawback, and so they too are increasingly amenable to wider forms of collaboration. Finally, for national policy communities, the significance of the EU as a reference point for new initiatives is largely instrumental, in that the EU is simply the international agency with which they are most frequently involved; there is no in-principle objection to a more diverse range of international dialogues and co-operative efforts.

In conclusion, while it seems highly likely that these trends towards the diversification and fragmentation of political authority will continue and indeed accelerate, it is very difficult to predict what their implications will be for the long-term future of European police co-operation. What we can be sure of, however, is that in a post-sovereign Europe, the maps with which we depict and comprehend the exercise of political authority, including authority over policing, will require frequent alteration and a more sophisticated interpretive key. And in this increasingly fluid scenario not only may the project of international police co-operation seem to lack any apparent final destination or ultimate purpose, but in presenting a moving target it may exacerbate the lack of transparency and deficient accountability which plague the current arrangements.

Notes

* The timetable for the production of this volume precludes acknowledgement in the text of the significant changes to the legal structure of European police co-operation introduced by the Treaty of Amsterdam, which was signed in October 1997 and awaits ratification. The Treaty of Amsterdam introduces into European law a new chapter on the so-called 'area of freedom, security and justice', within which are to be found both a new Title on free movement, immigration and asylum to be added to the EC Treaty and a revised and streamlined Title V1 of the Treaty on European Union covering police and customs co-operation and judicial co-operation in criminal matters. Superimposed upon this new dual structure is the entire Schengen system, now formally incorporated into the EU framework.

Arguably, the new arrangements do little to improve the coherence or account-ability of the system of police co-operation. The role of the Community institutions in the new Third Pillar is somewhat stengthened, but the overall developmental strategy has been at the cost of uniformity and transparency. The new dual structure, combined with the incorporation of Schengen and a generally more permissive approach to variable geometry, promises a much more fragmented approach to internal security questions in Europe and threatens to blur lines of responsibilty and answerability.

A much more detailed analysis of these developments, which relies on many of the ideas developed in the present chapter, is to be found in Walker (forthcoming).

1. Benyon *et al.* (1993); Spencer (1995); Anderson *et al.* (1995).
2. Anderson *et al.* (1995), Chapter 4.
3. *Statewatch*, Volume 6, No. 3 (May–June 1996), pp. 19–21.
4. Anderson *et al.* (1995), Chapters 6 and 8.
5. Under Article K.3(c) conventions which require to be adopted by Member States in accordance with their own constitutional requirements are established as the most solemn and binding form of legal agreement within the Third Pillar. Article K.3 also refers to two less-formal types of agreement, namely joint actions and joint positions.
6. The only area of competence mentioned under the initial objectives of Europol as set out in Article 2 of the Convention which does not also appear in the extended remit of the EDU is 'trade in human beings'.
7. *Statewatch*, Volume 6, No. 2 (March–April 1996), p. 21.
8. Walker (1996b).
9. Cullen (1992).
10. de Burca (1996); Walker (1996a).
11. The British policy of non co-operation was most evident in the JHA sphere at the meeting of the Council of Justice and Home Affairs Ministers in Luxembourg on 4 June 1996, where the British Home Secretary, Michael Howard, refused to agree to any decisions requiring unanimity. See *Statewatch*, Volume 6, No. 3 (May–June 1996), pp. 18–19.
12. Walker (1994a); Guyomarch (1995).
13. Marenin (1982), p. 258.
14. Milward (1992).
15. HM Government (1996).
16. Lindberg (1963); Pentland (1973).
17. Levi (1991); Lavoie (1992).
18. Cohen (1985), p. 37.
19. Proposal for a Council Directive on the elimination of controls on persons crossing internal frontiers, Com (1995) 347 final.
20. Anderson *et al.* (1995), Chapter 1.
21. Sheptycki (1995).
22. Holdaway (1983).
23. Weiler (1991), pp. 2476–7.
24. Weiler (1991); Peters (1992).
25. Bigo (1994), (1996); den Boer (1994); Waever (1996).

26. Smith (1991), Chapter 7.
27. Waever and Kelstrup (1993), p. 65.
28. Cable (1994).
29. Waever (1996), p. 122.
30. Anderson *et al.* (1995), Chapter 5.
31. Waever (1996), pp. 106–7.
32. Bigo (1994), p. 164.
33. Pearson (1983).
34. Waever (1996), p. 104.
35. den Boer and Walker (1993); Anderson (1994).
36. Benyon *et al.* (1993).
37. Schengen and Interpol have proved particularly resilient competitors to Third Pillar institutions within the European security market. Schengen should be redundant once a system of open internal frontiers and adequate compensatory security measures is in place within the EU as a whole but, as the robust rejection by Schengen's defenders of proposals to the 1996 IGC that its provisions should now be subsumed under the Third Pillar indicate, that point has not yet been reached (see *Statewatch*, Volume 6, No. 4 (July–August 1996), pp. 17–19). For its part, Interpol has concentrated over the past decade on improving the efficiency of its European activities, particularly through its Automated Search Facility which is intended to allow local police units across Europe to search and download information kept within Interpol's central database. See, generally, Anderson *et al.* (1995), Chapter 2.
38. Richardson and Jordan (1979).
39. Walker (1994b).
40. Garland (1996).
41. e.g. McBarnet (1981).
42. e.g. Crozier (1964).
43. Anderson *et al.* (1995), Chapter 2.
44. Storbeck (1996).
45. Anderson *et al.* (1995), Chapter 8.
46. Walker (1997).
47. Harmsen (1994); Walker (1996a).
48. *Statewatch*, Volume 6, No. 1 (January–February 1996), pp. 20–3 and Volume 6, No. 2 (March–April 1996), pp. 19–20.

Bibliography

Anderson, M. (1994) 'The Agenda for Police Cooperation', in M. Anderson and M. den Boer (eds), *Policing Across National Boundaries*, London: Pinter, pp. 3–21.

Anderson, M., den Boer, M., Cullen, P., Gilmore, W., Raab, C. D. and Walker, N. (1995) *Policing the European Union: Theory, Law and Practice*, Oxford: Clarendon Press.

Benyon, J., Turnbull, L., Willis, A., Woodward, R. and Beck, A. (1993) *Police Co-operation in Europe: An Investigation*, Leicester: Centre for the Study of Public Order.

Bigo, D. (1994) 'The European Internal Security Field: Stakes and Rivalries in the Newly Developing Area of Police Intervention', in M. Anderson and M. den Boer (eds) *Policing Across National Boundaries*, London: Pinter, pp. 161–73.

Bigo, D. (1996) *Police en Reseaux: l'experience européene*, Paris: Presses de Sciences Po.

Cable, V. (1994) *The World's New Fissures*, London: Demos.

Cohen, H. (1985) 'Authority: The Limits of Discretion', in F. A. Elliston and M. Feldberg (eds) *Moral Issues in Police Work*, New Jersey: Rowman & Allanheld, pp. 27–42.

Crozier, M. (1964) *The Bureaucratic Phenomenon*, London: Tavistock.

Cullen, P. (1992) *The German Police and European Co-operation*, Edinburgh: European Police Co-operation Working Paper No. 2.

de Burca, G. (1996) 'The Quest for Legitimacy in the European Union', *Modern Law Review*, **59**, pp. 349–76.

den Boer, M. (1994) 'The Quest for European Policing: Rhetoric and Justification in a Disorderly Debate', in M. Anderson and M. den Boer (eds) *Policing Across National Boundaries*, London: Pinter, pp. 174–96.

den Boer, M. and Walker, N. (1993) 'European Policing after 1992', *Journal of Common Market Studies*, **31**, pp. 3–28.

Garland, D. (1996) 'The Limits of the Sovereign State: Strategies of Crime Control in Contemporary Society', *British Journal of Criminology*, **36**, pp. 445–71.

Guyomarch, A. (1995) 'Problems and Prospects for European Police Co-operation after Maastricht', *Policing and Society*, **5**, pp. 249–61.

Harmsen, R. (1994) 'A European Union of Variable Geometry: Problems and Perspectives', *Northern Ireland Legal Quarterly*, **45**, pp. 109–33.

HM Government (1996) *A Partnership of Nations: The British Approach to the European Union Intergovernmental Conference 1996*, London: HMSO.

Holdaway, S. (1983) *Inside the British Police*, Oxford: Blackwell.

Lavoie, C. (1992) 'The Investigative Powers of the Commission with Respect to Business Secrets under Community Competition Rules', *European Law Review*, **12**, pp. 20–40.

Levi, M. (1991) 'Pecunia non olet: Cleansing the Money-Lenders from the Temple', *Crime, Law and Social Change*, **16**, pp. 217–302.

Lindberg, L. N. (1963) *The Political Dynamics of European Integration*, Stanford: Stanford University Press.

McBarnet, D. (1981) *Conviction*, London: Macmillan.

Marenin, O. (1982) 'Parking Tickets and Class Repression: the Concept of Policing in Critical Theories of Criminal Justice', *Contemporary Crises*, **6**, pp. 241–66.

Milward, A. S. (1992) *The European Rescue of the Nation-State*, Berkeley: University of California Press.

Pearson, G. (1983) *Hooligan: a History of Respectable Fears*, London: Macmillan.

Pentland, C. (1973) *International Theory and European Integration*, London: Faber.

Peters, B. G. (1992) 'Bureaucratic Politics and the Institutions of the European Community', in A. M. Sbragia (ed.) *Europolitics: Institutions and Policymaking in the 'New' European Community*, Washington D.C.: The Brookings Institute, pp. 75–122.

Richardson, J. J. and Jordan, A. G. (1979) *Governing under Pressure: the Policy Process in a Post-Parliamentary Democracy*, Oxford: Martin Robertson.

Sheptycki, J. (1995) 'Transnational Policing and the Making of a Postmodern State', *British Journal of Criminology*, **35**, pp. 613–35.

Smith, A. D. (1991) *National Identity*, London: Penguin.

Spencer, M. (1995) *States of Injustice: a Guide to Human Rights and Civil Liberties in the European Union*, London: Pluto.

Storbeck, J. (1996) 'Part of the Union', *Policing Today*, Volume 2, Part 1, pp. 28–31.

Waever, O. (1996) 'European Security Identities', *Journal of Common Market Studies*, **34**, pp. 103–32.

Waever, O. and Kelstrup, M. (1993) 'Europe and its Nations: Political and Cultural Identities', in O. Waever, B. Buzan, M. Kelstrup and P. Lemaitre (eds) *Identity, Migration and the New Security Agenda in Europe*, London: Pinter, pp. 40–92.

Walker, N. (1994a) 'European Integration and European Policing', in M. Anderson and M. den Boer (eds) *Policing Across National Boundaries*, London: Pinter, pp. 22–45.

Walker, N. (1994b) 'Reshaping the British Police: The International Angle', *Strategic Government*, **2**, pp. 25–34.

Walker, N. (1996a) 'European Constitutionalism and European Integration', *Public Law*, pp. 266–90.

Walker, N. (1996b) 'European Policing in Transition', in O. Marenin (ed.) *Policing Change, Changing Police: International Perspectives*, New York: Garland, pp. 251–84.

Walker, N. (1997) 'Deficient Weaponry, Reluctant Marksmen and Obscure Targets: Flaws in the Accountability of Undercover Policing in the EU', in M. den Boer (ed.) *Undercover Policing and Accountability from an International Perspective*, Maastricht: EIPA, pp. 205–16.

Walker, N. (forthcoming) 'European Policing and the Politics of Regulation', in P. Cullen and W. Gilmore (eds) *Crime sans frontières: International and European Legal Approaches*, Edinburgh: David Hume Institute.

Weiler, J. H. H. (1991) 'The Transformation of Europe', *Yale Law Journal*, **100**, pp. 2403–83.

Chapter 12

Frontier Issues before the European Court of Justice

PETER CULLEN

Introduction

European Community law is concerned almost by definition with frontier or border issues. The application of Community law is usually conditional upon there being some 'cross-border' issue to regulate or adjudicate upon; matters which are purely 'internal' to a Member State do not generally give rise to a Community law issue.[1] This chapter will examine the involvement of the European Court of Justice in frontier issues, with particular reference to the free movement of persons in the context of the internal market, a supposedly 'border-free' area. The first part of the chapter will examine the role played by the Court in developing the economic rights of free movement which were an integral part of the common market idea. The second section will examine the impact of the Court's jurisprudence on attempts by the Member States to limit access by Community nationals to their territory on public policy or public security grounds. The Court has allowed Member States only rather narrow scope to close their borders on these grounds. Third, the chapter will discuss a number of the issues arising for judicial resolution as a result of the completion of the internal market according to Article 7a of the EC Treaty. The chapter will conclude by considering Court of Justice jurisdiction in the context of Title VI of the Maastricht Treaty, on Cooperation in Justice and Home Affairs, better known as the 'Third Pillar' of the European Union. The Third Pillar raises a host of new frontier problems for the European Union.

Securing economic freedoms

The Court of Justice has been instrumental in helping litigants to overcome barriers which Member States have placed in the way of the exercise of 'Community rights'. Since its ground-breaking case law of the early 1960s concerning the 'direct effectiveness' of Community treaty provisions,[2] it has declared a large proportion of those provisions, in particular those concerning the abolition of restrictions to free trade or the free movement of persons, to be capable of producing rights which individuals or legal persons may claim before their national courts.[3] The principle of direct effects has opened the gates to many successful challenges to Member States' attempts to secure their frontiers against unwanted competition from foreign goods or to close them to nationals of other Member States seeking to pursue legitimate economic activities. In this section we will examine the Court's role in interpreting the EC Treaty provisions concerning the free movement of persons. The Treaty distinguishes between workers and the self-employed, with the latter category being sub-divided between persons who wish to establish themselves on a more or less permanent basis in another Member State, and the providers or recipients of services, seeking to carry out a more transitory economic activity.[4]

Common to these EC Treaty provisions on persons is a requirement that Member States abolish restrictions on the exercise of the freedom in question. The most blatant form of such restrictions is discrimination on the ground of nationality; both in its general and more specific provisions the Treaty prohibits such discrimination.[5] In this respect, the Treaty provisions on the free movement of persons are based on common principles.[6] The Court of Justice has elaborated on the free movement of persons provisions in a number of important ways. In particular, it has held that non-discriminatory restrictions on the free movement of persons may breach the EC Treaty and require to be justified on strict criteria if they are not to be prohibited. This approach concentrates on identifying obstacles to the exercise of Community freedoms; it recognizes that hindrances to the movement of Community nationals, or to their provision of cross-frontier services, may result from national rules which, on their face, apply equally to the domestic and foreign contexts. For example, in the recent much-discussed *Bosman* case, football association rules which restricted the movement of a soccer player from Belgium, who was out of contract, to another Member State, by imposing a transfer fee requirement, were found to breach the EC Treaty rules regarding free movement of workers (Article 48, EC Treaty), even though the rules applied equally to movement between clubs within Belgium.[7] Drawing on jurisprudence from the areas of the freedom of establishment and the freedom to provide services, the Court of Justice held that the restrictions constituted an unjustified limitation on the fundamental Community principle of free movement. There were no 'pressing reasons of public interest' to justify restricting M. Bosman's access to the employment market of another Member State. The case is especially significant because it represents further evidence of a convergence of the Court's treatment of national legal obstacles to the EC Treaty's economic freedoms. The Advocate-General's opinion in *Bosman* describes the trend thus:

in examining the compatibility of national provisions with the provisions of Community law on the fundamental freedoms, it is not so important which specific fundamental freedom a particular factual situation is to be measured against. What should be decisive is rather whether the provisions in question hinder *transfrontier economic activity* and – if that is the case – whether those restrictions are justified.[8]

The 'transfrontier economic activity' to which the Advocate-General refers has been further facilitated by the broad interpretation given by the European Court to the range of accompanying or 'flanking' rights conferred by Community law upon Community nationals and their families who exercise the right to move to another Member State. This is especially apparent with regard to the movement of Community workers. Article 48 of the EC Treaty sets out four particular elements of the right to free movement for workers: the right to 'accept offers of employment actually made'; 'to move freely within the territory of the Member States for this purpose', that is, for the purpose of accepting an offer of employment; 'to stay in a Member State for the purpose of employment' under the same conditions as nationals of that state; and finally the right to remain in the Member State where one has worked after one's employment there has ceased. This list was, however, declared by the European Court of Justice to be 'non-exhaustive', in a judgment which held that the Treaty rights to 'accept offers of employment actually made' and 'stay in a Member State for the purpose of employment' should not be interpreted strictly so as to prevent a national of a Member State from going to another Member State and staying there for a reasonable time in order to *seek work* on the spot.[9]

The willingness of the Court in this case to go beyond the precise wording of the Treaty was motivated both by the fundamental nature of the freedom in question – 'freedom of movement for workers forms one of the foundations of the Community' – and by the desire to 'make that provision [Article 48(3)] effective'.[10] This approach has attracted criticism from some quarters for producing legal uncertainty.[11] So consistent has the Court been in favouring a broad interpretation to rights connected with the exercise of freedom of movement that one could, however, argue that it was predictable that such an interpretation would be given. Luxembourg's kind-heartedness on these matters is clearly evident in its approach to the interpretation of Council Regulation 1612/68,[12] in particular its provisions on social and tax advantages and the educational rights of migrant workers' children.[13] While it is true that the Court has developed the law in these fields beyond the express Treaty text, it has undoubtedly made the right to free movement more real and effective in the process; its approach is consistent with the EC Treaty's objective of creating 'an internal market characterised by the abolition of obstacles to the free movement of persons' (Article 3(c)). There have certainly been more questionable examples of policy making by the Court.[14]

Public policy exclusions – the judicial response

As we have seen, the EC Treaty, as generously interpreted by the European Court of Justice, encourages trade and movement of Community nationals for economic purposes across frontiers. But it also allows these economic activities to be restricted, or stopped altogether, in order to protect certain competing interests, including those of public morality, public policy or public security.[15] Furthermore, Community law leaves to Member State jurisdiction the measures which may be taken in relation to persons not falling within the scope of its protection, with the result that the frontier may, and usually does, 'behave differently' towards the favoured nationals of Member States and the less-welcome nationals of so-called 'third countries'.[16] Within these limits, the Court of Justice has performed an extremely important role in controlling the way in which the Member States have used their law enforcement powers to restrict free movement across their frontiers. The focus for its jurisprudence has been the interpretation of the Treaty exceptions to free movement rights. One particularly interesting aspect of this jurisprudence is that it has led the Court into the territory of Member States' criminal jurisdiction, including the exercise of police powers. Although the criminal law has not been subjected to the harmonizing process undergone by Member States' economic laws, the imperative of fulfilling Community treaty objectives has nevertheless also resulted in impingement on national criminal jurisdiction.[17]

It would have been inconsistent with its broad interpretation of the Treaty's fundamental freedoms for the Court to adopt a similarly broad interpretation of the leeway the EC Treaty offers the Member States to derogate from these freedoms. The Court has interpreted the derogations strictly, guided in the field of the free movement of persons by an important Directive which regulates the exercise by Member States of the public policy and other provisos (Directive 64/221).[18] One of the aims of the Directive was to prevent the possible abuse by Member States of the derogations.[19] The Directive was equally important from the viewpoint of securing some measure of uniformity in the policies and procedures of Member States in relation to the exclusion of Community nationals from their territory. It insists that any measures of public policy, public security or public health which have the effect of frustrating Community rights – refusal of entry, refusal of a residence permit or expulsion from the territory – must 'not be invoked to service economic ends' and must 'be based exclusively on the personal conduct of the individual concerned'. Article 3(2) of the Directive stipulates that 'Previous criminal convictions shall not in themselves constitute grounds for the taking of such measures.' Directive 64/221 also secures some measure of 'approximation' of the procedural rules of the Member States; it provides for a 'double safeguard' against unfair exclusions, in the form of compulsory notification of the grounds of exclusion and a right of appeal for the person concerned.[20]

The terms of the Directive have been the subject of a significant number of cases before the European Court of Justice, frequently on reference from national criminal courts under the Article 177 preliminary rulings procedure. Most of the litigation has arisen from instances in which Member States have sought to invoke the public policy proviso to justify the denial of a Community law right of residence, though public policy claims frequently also encompass a public security element.[21] The Court has guided national courts and administrations, including immigration offices and police forces, to the type of behaviour which may legitimately attract the sanctions of refusal of entry, refusal of a residence permit or expulsion (deportation) from the national territory. The Luxembourg court has stated that

> the concept of public policy presupposes, in any event, the existence, in addition to the perturbation of the social order which any infringement of the law involves, of a genuine and sufficiently serious threat to the requirements of public policy affecting one of the fundamental interests of society.[22]

The Court has found support for this test in provisions of the European Convention for the Protection of Human Rights and Fundamental Freedoms of 4 November 1950 (ECHR) which allow derogations to Convention rights only in so far as they are necessary for the protection of interests in a democratic society.[23] If the illegal conduct concerned is sufficiently serious to fall within the Court's formula, the Member States may use the measures of expulsion or refusal of entry or continued residence referred to in Directive 64/221.

Thus, a general 'proportionality' test applies to the use of national sanctions which would frustrate the achievement of free movement. This general test was formulated in a case concerning relatively minor breaches of British anti-drugs legislation.[24] Its stiffness indicates to national authorities that only more serious criminal behaviour warrants a measure denying free movement. The European Court employs the proportionality principle consistently in the field of the free movement of persons to ensure that Member States do not unreasonably impede cross-frontier economic activity.[25] Further manifestations of the principle may be found in the rule that national authorities may adopt public policy or security measures under the EC Treaty derogations only on the basis of the individual position of the person protected by Community law, not on the basis of general considerations.[26]

And the existence of previous criminal convictions may be regarded as a sufficient, and therefore proportionate, ground for refusal of a Community right to enter or reside 'only in so far as the circumstances which gave rise to [the] conviction are evidence of personal conduct constituting a present threat to the requirements of public policy'.[27]

New challenges for the European Court post-1992

The case law discussed so far demonstrates that the European Court had already made a big contribution to the achievement of the Treaty objective of free movement of persons before the Community decided to 'complete' the creation of the internal market by 31 December 1992, that is, the date set out in Article 8a of the EEC Treaty, by creating an area without internal frontiers. This article was introduced into the EEC Treaty by the Single European Act; it survived the Maastricht revisions intact except for being renumbered Article 7a of the EC Treaty. The 1985 Commission White Paper, whose thinking underpinned the new article, envisaged a number of measures in the field of free movement of persons to complete the work hitherto undertaken by the Court and the Community's legislative institutions. The White Paper was unambiguous in its call to Member States to 'get rid entirely' of the physical and other barriers which remained between them, on the ground that their continued existence ran counter to the idea of a single market. The removal of physical barriers – in particular customs posts, being the 'most visible sign of the continued division of Europe' – would be 'the clearest sign of the integration of the Community into a single market'.[28] This blueprint had obvious implications for the free movement of persons, the Commission noting that the removal of checks on persons crossing intra-Community borders would require the development of 'alternative ways of dealing with other relevant problems such as public security, immigration and drugs controls'.[29] Reconciling the unhindered mobility of persons with the preservation of a maximum degree of security within the single market area has been a constant concern of European policy makers since the entry into force of Article 8a of the EEC Treaty. The Member States gave expression to this tension when they appended to the internal market provisions of the Single Act a general declaration asserting that nothing in the Act would affect their rights to combat crime and control immigration from third countries. Immigration from third countries into the Union is frequently bracketed by Member States alongside transfrontier crime as a 'security problem', to be addressed if necessary by criminal policy measures.[30]

The main solution offered by the Commission and the Member States to the 'mobility vs security' conundrum has been to strengthen the external border of the Community. There is a paradox here. The Commission has scorned the idea that Member States' internal frontiers represent an effective means of controlling cross-border criminal activity.[31] Yet at the same time it espouses the desirability of ever more stringent controls at the external frontiers of the European Union.[32] According to this 'logic', the border can indeed perform an effective security function, but it depends on the geography. What the Commission really wants, therefore, is not the end of borders as a security 'filter', but the repositioning, and indeed strengthening, of security (including immigration) controls at the external frontier of the European Union; it has been a consistent supporter of the Schengen treaty arrangements (Schengen Treaties of 1985 and 1990), the Dublin Asylum Convention (1990) and the draft External Frontiers Convention (1993),[33] all of which subscribe to this philosophy.

Unfortunately for the Commission, the clarity of Article 8a of the EEC Treaty left much to be desired; in particular, it was not entirely clear whether it required the complete abolition of internal border controls and, relatedly, whether it applied to controls carried out on Community nationals only or also nationals of third countries. There was also a question as to the binding quality of the 31 December 1992 deadline. And the distinction between frontier checks and police checks – the issue being whether the latter could be permitted if the former were not – was not resolved by the Treaty either. These questions could, of course, have been addressed by the Community implementing legislation and/or by legal action in the European Court of Justice, resulting in a uniform Community interpretation. The Court is entrusted by the EC Treaty with the task of providing uniform interpretations of Community law (see Articles 164 and 177). Clarificatory legislation was not proposed by the Commission until the middle of 1995, nearly three years after the 1992 deadline. This can be explained partially by the Commission's belief that Article 8a was in fact complete in itself. In a trenchant legal opinion of 8 May 1992, it appeared to provide definitive answers to the above questions: what was required was the 'complete abolition', by 31 December 1992, of controls on all persons crossing an internal border of the Community, 'irrespective of their nationality';[34] law enforcement powers which were exercisable over the whole of a Member State's territory would, however, continue to be permissible under Community law, even when carried out near the frontier.[35]

In order to reinforce its message to the Member States that Article 8a required complete abolition of border controls and allowed the Member States no 'margin of discretion', the Commission stated in its 1992 opinion that it would 'use all the legal and political means at its disposal' to ensure that the Member States adhered to the 31 December deadline.[36] This was a rather explicit threat to take to the European Court, under Article 169 'infringement proceedings', any Member State which failed to pursue the measures required to implement the treaty provision. It was widely assumed that the United Kingdom, in particular, was in the firing line for its declared unwillingness to abolish border controls entirely. The British government has always parted company with the Commission on the legal interpretation of Article 8a EEC (and now 7a EC). The view from London is that this provision of Community law does not prevent Member States from checking the nationality of persons, including Community nationals, on their entry into the UK (House of Commons, 1992). The commencement of Article 169 proceedings is, however, a notoriously sensitive matter and the Commission exercises a discretion whether or not to 'prosecute', in the light of political and other factors.[37] Commissioner Bangemann announced in September 1992 that the Commission had reached a compromise solution with the United Kingdom government and 'had no interest in a legal wrangle . . . over the abolition of border controls'.[38] This decision may well have been influenced by a desire not to antagonize the UK government during the debates on ratification of the Maastricht Treaty. The Court was, thus, deprived on this occasion of the opportunity to pronounce on the interpretation of Article 8a.

The Commission would not have had to bring forward legislation in the middle of 1995 to implement its open borders policy had there not been such lack of progress by Member States in adopting measures to compensate for the removal of controls. Uncertain of the Community's legal authority and anxious not to let matters of 'legal doctrine' interfere with 'practical effectiveness', the Commission went along with the decision of the Member States to tackle the compensation issue by intergovernmental, as opposed to the Community, decision-making framework.[39] The 'Palma Document', presented to the meeting of the European Council at Madrid in June 1989, established the priority areas for action by Member States in respect of external and internal frontiers respectively.[40] These areas included: definition of common procedures for checks at the external frontiers, improved co-operation between law enforcement agencies, combating illegal immigration networks and a variety of other 'off-setting measures', concerning *inter alia* drug trafficking, judicial co-operation, visa policy, right of asylum (including the determination of the state responsible for examining the asylum application), refugee status and the intensification of anti-terrorist co-operation.

It is highly doubtful that any of these matters could have been brought clearly within the scope of express EEC Treaty competence; there was also no likelihood of achieving the unanimous agreement required for use of the 'catch-all' Article 235 of the Treaty. It should be noted that the Commission's proposal of 12 July 1995 for a Council Directive on the elimination of controls on persons crossing internal frontiers does not claim Community competence over these matters; indeed the entry into force of the Directive is made expressly conditional upon adoption by the Member States and entry into force of the full range of 'accompanying measures essential to the elimination of controls'.[41] The Directive's aim is to confirm that the Member States are under an obligation, arising from Article 7a of the EC Treaty, to abolish completely controls on all persons, whatever their nationality, crossing intra-Community frontiers (Article 1). The European Parliament, whom the 12 July proposal is designed to mollify, claimed that Article 8a EEC (now 7a EC) conferred upon the Commission the right and indeed the obligation to bring forward *Community* proposals to give effect to the free movement of persons by 31 December 1992. So incensed was it by the Commission's failure to do so that in November 1993 it lodged an action with the European Court of Justice claiming that the Commission was in breach of its Treaty obligations.[42] The Parliament's statement of claim to the Court referred to a mistaken Commission policy of

> acquiesce[nce] in the Member States' wish to act outside the framework set by the [EC] Treaty and to conclude intergovernmental agreements among themselves.

The Parliament, in other words, believes that it is for Community law to regulate compensation issues.

Commenting on the 1995 proposals, some Strasbourg parliamentarians rejected the Commission's continued linkage of abolition with compensation and suggested that the removal of frontier controls should proceed immediately and independently of the latter.[43] This view reflects frustration with the continuing delays surrounding adoption or ratification by Member States of the outstanding compensatory instruments, including, notably, the Dublin Asylum Convention and the External Borders Convention.[44] In the light of the proposals made by the European Commission in July 1995 to expedite the elimination of intra-Community frontier controls and to grant a limited 'right to travel' to third country nationals resident within the European Union, the European Parliament as a whole decided, however, in 1996 not to pursue its action against the Commission.[45] Thus the European Court was again deprived of the opportunity to offer an interpretation of Article 7a of the EC Treaty. There is, however, case law of the Court concerning the legitimacy under Community law of border and security controls on Community nationals carried out at an internal Community frontier. Although the case law arose before the date set for completion of the internal market, the Court's decisions, reached in cases brought by the Commission against Belgium and the Netherlands respectively, offer useful insights into the likely approach of the Court to controls applied after 31 December 1992.[46]

In *Commission v Belgium* the European Court decided that it was not incompatible with Community law for Belgian law enforcement officials to inspect residence or establishment permits held by non-Belgian Community nationals on their entry into Belgium. These inspections were carried out throughout Belgian territory, on a non-systematic basis, as part of a general system of police checks, to which all inhabitants were subject. The Court's opinion stressed the fact that they did not constitute a 'condition of entry into Belgian territory'; failure to produce such a card would result in a fine, not denial of entry.[47] The Court recognized, however, that controls of this kind could 'constitute a barrier to the free movement of persons within the Community, a fundamental principle of the EEC Treaty' if 'carried out in a systematic, arbitrary or unnecessarily restrictive way'.[48] In the Dutch case, the Court of Justice held that the Netherlands had broken Community rules on free movement by requiring Community nationals who wished to enter Dutch territory to answer questions put routinely by border officials regarding the purpose and duration of their journey and the financial means at their disposal. It was not consistent with Community law to establish any conditions of entry other than the production of a valid passport or identity card.[49] The United Kingdom, intervening in the proceedings, wanted the Court to rule whether questions asked for reasons of public policy or public security would contravene Community law. The Court declined to deal with the point as it was not relevant, but the Advocate-General suggested that it would always be permissible for frontier officials to

> put questions to people whose behaviour is such as to raise suspicions or, in any event, in circumstances in which public security appears particularly threatened.[50]

Any information requested from and measures taken against individuals on public policy grounds would nevertheless have to 'be justified by the existence of particular circumstances' and satisfy the principles of the Court's case law on public security issues.[51]

The Commission clearly had regard to this case law in drafting the proposal of 12 July 1995 on the elimination of frontier controls (Commission, 1995a). Article 1(2) of the draft, for example, states explicitly that the abolition of controls on persons crossing an internal Community frontier will not affect the exercise of law enforcement checks carried out across the whole of a Member State's territory. Also, the proposal does not seek to prevent national laws from requiring persons to carry documents such as driving licences or identity cards; by implication, a general system of police checks of such documents would not breach Community law. Controls of the type applied by Belgium in the case described above could, thus, continue to be carried out, but the Commission would presumably insist that Member

States observed the Court's proportionality test, that is, the checks should be sporadic and targeted, rather than systematic, and must be fully justified by the specific circumstances of the case. The Commission also relies on the Court's jurisprudence in stressing that such checks would have to be exercised 'without discrimination between domestic and cross-border traffic'.[52] Article 2 of the proposal permits Member States to reinstate frontier controls for thirty-day periods 'in the event of a serious threat to public policy or public security'. But it also insists on respect for the proportionality principle in this context: the reinstated controls 'shall not exceed what is strictly necessary to respond to the serious threat'. This draft legislation, should it enter into force, would of course be subject to the jurisdiction of the Court of Justice. References from national courts for interpretation of the Directive's provisions concerning 'law enforcement powers' (not defined) or 'serious threat to public policy' (not defined) could be expected. The Court would most probably draw on its jurisprudence concerning public policy restrictions to assist it in this task (see above).

Luxembourg has not itself provided an interpretation of Article 7a in preliminary ruling or any other proceedings. It appeared likely that a reference would be made by the English High Court in the case of Donald Flynn. Mr Flynn, a British citizen, refused to show his passport to a British Immigration Officer at Dover after returning in May 1993 from a day-trip to Calais.[53] He claimed that, after 31 December 1992, such passport controls were outlawed by Article 7a EC. The High Court took the view that the article did not 'ordain' that the internal market must come into effect on 1 January 1993. The English court believed that the UK was, therefore, justified in maintaining passport controls on Community nationals entering its territory. The judge noted in his conclusion that the controls applied in the UK to Community nationals were 'very light';[54] he did not, however, address the proportionality issue directly. It is surprising that the High Court did not refer this case to Luxembourg, given that the judge recognized there were a number of open questions concerning the interpretation of Article 7a, in relation to which the view of the European Court could only be guessed at.[55]

Maastricht's Third Pillar – shutting out the Court from border issues

Unless its jurisdiction is specifically invoked by a Convention adopted under Article K.3, paragraph 2(c), of the Maastricht Treaty, the European Court of Justice is excluded from reviewing action taken by the Member States under the Maastricht Treaty provisions concerning Cooperation in Justice and Home Affairs. The Treaty confines mandatory Court of Justice jurisdiction to

First Pillar, that is, Community law, matters. The Member States decided at their Maastricht Summit in 1991 that there should only be a 'permissive' role for the Court in relation to the 'matters of common interest' listed in Article K.1, such as asylum policy, control of the external borders of the European Union and immigration policy, policy regarding nationals of third countries and European police co-operation. As well as being matters with a strong cross-frontier dimension, the substantive areas chosen for inclusion in Title VI are particularly sensitive ones for Member States: control over immigration and policing powers are jealously guarded items of national public policy.

Commentators were quick to recognize the overlaps between First and Third Pillar decision-making on free movement of persons issues, for example in the field of visa policy, in relation to which the Maastricht Treaty introduced Community law provisions (Article 100c EC Treaty) as well as intergovernmental ones.[56] An obvious functional connection between the Pillars is provided by the compensation objective, which was discussed in the previous section: Article K.1 of the Maastricht Treaty confirms that the primary objective of Title VI is to help achieve the free movement of persons. The supposedly complementary nature of the Pillars cannot, however, be taken for granted: for example, the Commission's proposal for a right to travel for third country nationals resident in a Member State sits uneasily with the generally restrictive approach to policy on third country nationals followed by Member States in Third Pillar negotiations.[57] It falls to the Commission, in the main, to secure the coherence of policy across the Union. One cannot, however, exclude the possibility that the Court of Justice may make a ruling relating to action under the Community Pillar which encroaches on a matter assumed to fall within Title VI intergovernmental cooperation.[58] Article M of the Maastricht Treaty indicates that legal conflicts between the Pillars are to be resolved in favour of Community powers; Article K.1 reinforces the point in connection with the Third Pillar.

The exclusion of the Court of Justice from review of Third Pillar decision-making has been strongly criticized on the ground that it may lead to a diminution in the protection of human rights.[59] Human rights considerations are specifically alluded to in Article K.2(1) of the Maastricht Treaty, which requires that the matters of common interest set out in Article K.1 be dealt with in compliance with the ECHR and the 1951 Geneva Convention relating to the Status of Refugees, and with regard to the protection offered by Member States to persons prosecuted on political grounds. The Member States must, therefore, respect the rights under these international law instruments of asylum seekers and other categories of would-be immigrants to the European Union, when taking measures pursuant to Title VI. These are the persons who are particularly exposed to the coercive power of the state in this context.[60] Clearly, if the Article K.2(1) provision is to benefit such persons, it must be capable of enforcement before a judicial tribunal; it should be noted, however, that it is not enforceable as a norm of Community law. The main effect of Article K.2(1) would seem to be to confirm international law obligations of the Member States, though the phrase 'protection afforded by Member States to persons persecuted on political grounds' would also encompass the protection afforded such persons under existing national (including constitutional) law. In so far as the international law obligations have effect within Member States' national legal orders, these obligations would be enforceable before national courts. This would, for example, enable ECHR provisions to be applied at national level in a majority of Member States; relief may eventually be sought from the Commission or Court of Human Rights in Strasbourg by litigants in any of the Member States. Twomey has explained how the ECHR, as interpreted by the Strasbourg Court, may be of particular value to persons subject to Third Pillar measures.[61]

Thus, we can see that an element of national and even supranational judicial review, involving the protection of individual rights, may be introduced into Third Pillar matters, even without the involvement of the European Court of Justice. Without the ability to make a reference to the European Court under the Article 177 reference procedure, the national courts would not have the option of referring a question to Luxembourg for an authoritative ruling, say on the interpretation of the External Borders or Europol Convention. The Strasbourg system does not employ this procedure, and, in any case, litigants may be reluctant to engage in the cost and delay of proceedings before the Commission or Court of Human Rights, even in cases where their claims may be admissible. It is quite foreseeable that national courts might interpret the provisions of a particular Third Pillar Convention in different ways, according to their divergent national human rights standards. This would mean that such Conventions would not be applied and could not be enforced uniformly throughout the European Union without Court of Justice jurisdiction, a point noted by the House of Lords Select Committee in its opinion concerning the Europol Convention.[62] The application within the field of Third Pillar action of national standards regarding the protection of human rights may of course be welcomed by those who regard the jurisprudence of the European Court of Justice concerning the general principles of Community law to be comparatively deficient.[63] The Court certainly, however, relies on ECHR provisions as part of these principles and could be expected, therefore, carefully to balance the desire of the Member States to curtail immigration or asylum to the European Union with the interests of the individuals concerned. As we have seen, it conducts a similar balancing exercise in its jurisprudence concerning the free movement of persons subject to Community law.

The Court might face a flood of cases once the Amsterdam Treaty brings asylum and immigration within the Community remit.[64]

Adoption of the Third Pillar, the decision of the Member States to exclude the Court was increasingly called into question. The questioning came from the Member States themselves as much as from academic commentators or other observers. It took the form of intensive debates on the question whether, or to what extent, the Court's permissive jurisdiction should be invoked under Article K.3 with regard to Third Pillar Conventions. It is not known what weight the Member States attached to the permissive jurisdiction provisions of the Third Pillar when they were drafting the Maastricht Treaty. The decision to use the Convention as the main legal instrument for implementation of the Third Pillar has, however, enabled Member States who support Court jurisdiction, as well as similarly minded Community institutions, to raise the issue as a matter of course in negotiations on Third Pillar Conventions. The United Kingdom fiercely opposed a role for the Court in Title VI matters. In oral evidence to the House of Lords Select Committee which met to discuss the Europol Convention, the British Home Secretary stated that this was because the Government feared 'expansive interpretation' by the European Court.[65] The other Member States were, however, able by the middle of 1996 – a year after the Europol Convention itself was signed – to persuade the British to agree to a Protocol to the Convention, enabling any Member State to opt to invoke the Court's preliminary rulings jurisdiction in questions concerning the Convention. By deciding, predictably, not to 'opt in' to such jurisdiction, the interpretation and application of the Convention within the United Kingdom will be left to the national courts and police authorities. It is too early to say whether the 'Europol compromise' has established a precedent for subsequent Third Pillar Conventions, but the Florence European Council recommended to the Council of Ministers that a similar solution be considered for two other Conventions signed in July 1995.[66]

The operation of Title VI was reviewed by the Intergovernmental Conference whose work commenced in March 1996, and which resulted in the Amsterdam Treaty. The Third Pillar will be confined to provisions 'police and judicial co-operation in criminal matters'. The new article K7 of the Treaty on European Union will enable Member States to confer any rulings jurisdiction on the Court of Justice. In previous sections of this chapter we have seen how the Court interpreted, in an expansive fashion, the free movement rights of economically active persons. Its case law enhanced the realization of the internal market, dovetailing with Commission action to achieve the same purpose; the Court limited the exercise of Member States' discretion to derogate from the free movement principle. The Single European Act of 1986 was not, however, unambiguous in its promotion of a border-free Europe for persons. It did not, for example, make clear whether the obligation to abolish frontier controls was absolute or whether police controls at the frontier would continue to be permitted. The Maastricht Treaty, more explicitly than the Single European Act, endorsed the legitimacy and utility of frontier controls at the external borders of the Member States, without providing any clear means of legal protection for persons exposed to such controls. Perhaps it was anxiety about the effective and uniform application of Third Pillar Conventions, rather than any obvious desire to improve such protection, which forced a rethink on the general exclusion of the Court of Justice at Amsterdam. Member States should not, however, need to be reminded that the Court has conferred important rights on individuals in the context of its enforcement of their Community law obligations.[67] It may be predicted that it would use any jurisdiction it was given over the Third Pillar to do the same. The Amsterdam Treaty will present the Court with a new range of challenges over free movement of persons and the policing of the external frontier. Member States are likely to scrutinize very carefully the exercise by the Court of its new competences.

Notes

1. See, for example, Case 175/78 *R v Saunders* [1979] ECR 1129.
2. Case 26/62 *Algemene Transport Onderneming van Gend en Loos v Nederlandse Administratie der Belastingen* [1963] ECR 1.
3. Collins (1990), pp. 122–6.
4. See Articles 48–66 of the EC Treaty.
5. Compare Articles 6 and 48(2) of the EC Treaty.
6. Case 48/75 *Jean Noel Royer* [1976] ECR 497, at p. 509.
7. Case C-415/93 *Union Royale Belge des Societes de Football Association ASBL v Jean-Marc Bosman* [1995] ECR I-4921.
8. Case C-415/93 *Bosman* [1995] ECR I-4921, at p. I-5006 (author's italics).
9. Case C-292/89 *R v Immigration Appeal Tribunal ex parte Antonissen* [1991] ECR I-745, at pp. I-776–I-777.
10. *Ibid.*
11. Hartley (1996), pp. 278–80.
12. Regulation (EEC) No. 1612/68 of the Council of 15 October 1968 on Freedom of Movement for Workers within the Community, OJ. Sp. Ed. 1968, L 257/2, p. 475.
13. Green *et al.* (1991), pp. 150–4, 188–90.
14. Weatherill and Beaumont (1995), pp. 174–82.
15. See relevant EC Treaty provisions, including Article 36 (for goods), Article 56 (for persons).
16. Countries which are not members of the European Union or are not part of the 'European Economic Area' (see Edward and Lane [1995], p. 147). Third country nationals do not enjoy any general right to free movement under Community law (Plender [1995], p. 15).

17. Anderson *et al.* (1995), p. 216.
18. Council Directive of 25 February 1964 on the co-ordination of special measures concerning the movement and residence of foreign nationals which are justified on the grounds of public policy, public security or public health (64/221/EEC), OJ. Sp. Ed. 1964, No. 850/64, p. 117).
19. Green *et al.* (1991), p. 127.
20. See Articles 6–9 of Directive 64/221 and comment by the Court of Justice in Case 36/75 *Roland Rutili v Ministry of the Interior* [1975] ECR 1219, at p. 1233.
21. Arnull (1990), p. 93.
22. Case 30/77 *R v Pierre Bouchereau* [1977] ECR 1999, at p. 2014.
23. See, for example, Article 8 of the Convention concerning protection of family life, and comment by the Court of Justice in Case 36/75 *Rutili v Ministry of the Interior* [1975] ECR 1219, at p. 1232.
24. Case 30/77 *Bouchereau* [1977] ECR 1999.
25. Arnull (1990), pp. 70–113.
26. Case 48/75 *Royer* [1976] ECR 497, at pp. 514–15.
27. Case 30/77 *Bouchereau* [1977] ECR 1999, at pp. 2013–14.
28. Commission (1985), p. 6.
29. *Ibid.*
30. Busch (1995), p. 155.
31. Commission (1988), p. 10.
32. *Ibid.*, pp. 11–12.
33. Commission's 1993 re-draft (COM [93] 684, 10 December 1993) of the *Ad Hoc* Group on Immigration's confidential 1991 draft.
34. Commission (1992a), p. 12.
35. *Ibid.*, p. 9.
36. *Ibid.*, p. 6.
37. Hartley (1994), p. 313.
38. Commission (1992b).
39. Commission (1995a), pp. 5–6.
40. House of Lords (1989), pp. 55–64.
41. Commission (1995a), p. 5.
42. Case C-445/93, *European Parliament v Commission*, OJ C 1, 4 January 1994, 12–13.
43. *Das Parlament* (2 August 1996), p. 9.
44. D'Oliveira (1994). The Dublin Convention finally entered into force only on 1 September 1997, having been signed in Dublin on 15 June 1990. OJ C254, 19 August 1997, pp. 1–12 (text) and OJ C268, 4 September 1997 (note on entry into force).
45. Case suspended 11 July 1996 (information supplied by Court of Justice).
46. O'Keeffe (1992), p. 11.
47. Case 321/87 *Commission v Belgium* [1989] ECR-I-997, at p. 1011.
48. *Ibid.*
49. Case 68/89 *Commission v Netherlands* [1991] ECR-I-2637, at p. 2655.
50. *Ibid.*, at pp. I-2648–I-2649.
51. Case 68/89 *Commission v Netherlands* [1991] ECR-I-2637, at p. I-2649.
52. Commission (1995a), p. 7.

53. *R v Secretary of State for the Home Department ex parte Donald Walter Flynn* [1995] 3 CMLR 397.
54. *Ibid.*, at p. 415.
55. One of these points is the crucial question of the linkage, as a matter of law, between any supposed obligation to abolish frontier controls and the adoption, by Community or Member States, of harmonizing compensatory measures: see the Judge's reference, at pp. 413–14, to Case C-297/92 *Istituo nazionale della previdenza sociale (INPS) v Corradina Baglieri* [1993] ECR I-5211, which addresses this point somewhat obliquely in the context of social security legislation (p. I-5233).
56. Hailbronner (1994), p. 995.
57. Cf. Commission (1995b).
58. Anderson *et al.* (1995), p. 210.
59. O'Keeffe (1995), p. 910; Twomey (1995).
60. Twomey (1995), p. 52.
61. Twomey (1995).
62. House of Lords (1995), pp. 30–1. 'Uniformity' is of course a relative concept. The implementation of judgments of the European Court, and of Community law generally, is subject to a number of national variables which make the achievement of full uniformity of legal application a near impossibility (Daintith [1995]).
63. Rengeling (1993), pp. 165–6.
64. Plender (1995), pp. 49–50.
65. House of Lords (1995), Evidence, 95.
66. Commission (1996), p. 12.
67. See case referred to in note 2, above.

Bibliography

Anderson, M., den Boer, M., Cullen, P, Gilmore, W. C., Raab, C. and Walker, N. (1995) *Policing the European Union. Theory, Law and Practice*, Oxford: Clarendon Press.

Arnull, A. (1990) *The General Principles of EEC Law and the Individual*, London: Leicester University Press.

Busch, H. (1995) *Grenzenlose Polizei. Neue Grenzen und polizeiliche Zusammenarbeit in Europa*, Münster: Westfälisches Dampfboot.

Collins, L. (1990) *European Community Law in the United Kingdom* (fourth edition), London: Butterworths.

Commission of the European Communities (1985) *Completing the Internal Market. White Paper from the Commission to the European Council* (Milan, 28 and 29 June 1985), COM (85) 310 final, 14.6.1985.

Commission (1988) *Communication of the Commission on the Abolition of Controls of Persons at Intra-Community Frontiers*, COM (88) 640 final, Brussels, 7.12.1988.

Commission (1992a) *Abolition of Border Controls, Commission Communication to the Council and to Parliament*, SEC (92) 877 final, Brussels, 8.05.1992.

Commission (1992b) *The Week in Europe*, WE 31/92, 10 September 1992 (London Office).

Commission (1995a) *Proposal for a Council Directive on the Elimination of Controls on Persons Crossing Internal Frontiers*, Brussels, COM (95) 347 final, 12.07.1995.

Commission (1995b) *Proposal for a Council Directive on the Right of Third Country Nationals to Travel in the Community*, Brussels, COM (95) 346 final, 12.07.1995.

Commission (1996) *Bulletin of the European Union*, No. 6/96, Brussels and Luxembourg: Office for Official Publications of the EC.

Daintith, T. (ed.) (1995) *Implementing EC Law in the UK: Structures for Indirect Rule*, Chichester: Wiley Chancery.

Das Parlament, 46(32), 2 August 1996.

D'Oliveira, H. U. J. (1994) 'Expanding External and Shrinking Internal Borders: Europe's Defence Mechanisms in the Areas of Free Movement, Immigration and Asylum', in D. O'Keeffe and P. M. Twomey (eds) *Legal Issues of the Maastricht Treaty*, London: Wiley Chancery.

Edward, D. A. O. and Lane, R. (1995) *European Community Law. An Introduction*, Edinburgh and London: Butterworths and Law Society of Scotland.

Green, N., Hartley, T. C. and Usher, J. A. (1991) *The Legal Foundations of the Single European Market*, Oxford: Oxford University Press.

Hailbronner, K. (1994) 'Visa Regulations and Third-Country Nationals in EC Law', *Common Market Law Review*, 31, pp. 969–95.

Hartley, T. C. (1994) *The Foundations of European Community Law* (second edition), Oxford: Oxford University Press.

Hartley, T. C. (1996) 'Five Forms of Uncertainty in European Community Law', *Cambridge Law Journal*, **55**(2), pp. 265–88.

House of Commons (1992) Select Committee on European Legislation, Second report, Session 1992–3, HC 79-ii, 24 June 1992, London: HMSO.

House of Lords (1989) Select Committee on the European Communities, Session 1988–9, 22nd Report, *1992: Border Control of People* (with Evidence), HL Paper 90, London: HMSO.

House of Lords (1995) Select Committee on the European Communities, Session 1994–5, 10th Report, *Europol* (with Evidence), HL Paper 51, London: HMSO.

O'Keeffe, D. (1992) 'The Free Movement of Persons and the Single Market', *European Law Review*, **17**, pp. 3–19.

O'Keeffe, D. (1995) 'Recasting the Third Pillar', *Common Market Law Review*, **32**, pp. 893–920.

Plender, R. (1995) *Immigration and Asylum Policies within the European Union*, Europa Institute, University of Edinburgh, Occasional Paper 5.

Rengeling, H.-W. (1993) *Grundrechtsschutz in der Europäischen Gemeinschaft*, Munich: Beck.

Twomey, P. (1995) 'Title VI of the Union Treaty: "Matters of Common Interest" as a Question of Human Rights', in J. Monar and R. Morgan (eds), *The Third Pillar of the European Union*, Brussels: European Interuniversity Press.

Weatherill, S. and Beaumont, P. (1995) *EC Law* (second edition), Harmondsworth: Penguin.

Chapter 13

The Protection of Cultures

JOSEPH A. McMAHON

Introduction

In February 1996, the Commission was warned by the President of the European Radio Association that unless action was taken against a French law requiring radio stations to devote at least 40 per cent of their broadcast time to songs in French a grave threat would arise to the integrity of the internal market. The Commission has decided not to take action. The French example has been followed by Ireland which has set a broadcasting quota of 30 per cent for a 'proper proportion of material of Irish origin and of Irish performance'. In this case legal action was started but has since been abandoned.[1] These two examples illustrate the nature of the clash between ensuring the integrity of the internal market, a Commission responsibility, and the protection of national culture, a Member State responsibility. They also illustrate an issue of frontiers, namely whether the creation of an internal market, and the resulting elimination of barriers to trade, will have an adverse impact on the protection of the various cultures of the Member States.

As a result of the Maastricht Treaty, the Community has now acquired a specific competence in relation to culture. Whereas the original Treaty of Rome contains few references to the protection of national culture, the Maastricht Treaty provides, in Article 3(p), that the Community shall contribute to the 'flowering of the cultures of the Member States'. This is further developed in Article 128 which states, *inter alia*, that:

> The Community shall contribute to the flowering of the cultures of the Member States, while respecting their national and regional diversity and at the same time bringing the common cultural heritage to the fore.

In subsequent paragraphs, Community action is to be directed at encouraging co-operation between the Member States and supporting and supplementing their action in four areas: the improvement of the knowledge and dissemination of the culture and history of the European peoples; the conservation and safeguarding of cultural heritage of European significance;

non-commercial cultural exchanges; and artistic and literary creation, including the audio-visual sector. The Council of Europe is specifically referred to in paragraph 3 as one of the competent international organizations with which the Community and the Member States are to foster co-operation. It is clear for these provisions that Community action in the sphere of culture is designed to supplement that of the Member States on whom the primary responsibility remains. This is modified to some extent by paragraph 4 of Article 128 which provides that: 'The Community shall take cultural aspects into account in its action under other provisions of this Treaty.'

It is in this context that the threat posed to the integrity of the single market by the French and Irish broadcasting laws and the threat posed to national cultures by the elimination of barriers to trade will have to be met. It is this area, the cultural aspects of other provisions of the Treaty and, in particular, the impact of culture on the free movement of goods provisions of the Treaty, which forms the basis for the remainder of this chapter.

The problem identified?

Two cases can be mentioned to highlight the nature of the problems which may arise through the application of the Treaty to areas of culture. Both of these cases involve books and the first of these cases, *VBVB & VBBB v Commission*, concerned the application of Article 85, the main competition provision of the Treaty.[2] The facts of the case are as follows: VBBB, a Dutch association of publishers, and VBVB, a Flemish association of publishers had notified the Commission in 1962 of an Agreement laying down rules relating to the book trade between the Netherlands and Flanders which they had concluded in 1949 and which had been amended in 1958. The provisions of that Agreement involved an exclusive dealing undertaking and a resale price mechanism. In 1978 the Commission sent their objections to the Agreement to the parties involved and in 1981 they decided that the Agreement was contrary to Article 85(1). The parties to the Agreement applied for a declaration that the Commission decision was void for reasons of procedural irregularities and of substance. In relation to the latter, the parties argued that the decision breached Article 10 of the European Convention on Human Rights, guaranteeing freedom of expression, Article 10 bis of the Paris Convention for the Protection of Industrial Property, and that Article 85(1) was not applicable.[3]

After concluding that there was an Agreement between associations of undertakings within the meaning of Article 85, the Court moved on to examine the substantive arguments of the parties.[4] On the issue involving Article 10 of the European Convention on Human Rights, it concluded that it had not been breached.[5] Article 10 bis of the Paris Convention was then analysed by the Court. It was argued that a system of resale price maintenance was necessary to prevent the practice of loss leading and, as a result, unfair competition. In response, the Court noted:

> The fact that a system of resale price maintenance may have the incidental effect of preventing unfair competition of the kind described by the applicants is not, however, a sufficient reason for failing to apply Article 85(1) to a whole sector of the market such as the book trade.[6]

There were other remedies for the abuses argued by the applicants and the fact that such abuses existed could not, therefore, justify the infringement of the Community rules on competition. The substantive arguments of the parties and the application were totally dismissed.

It is interesting to note that the Court assiduously avoided any mention of the cultural overtones of its decision. In its decision, the Commission had considered that cultural values could only be taken into account in a negative manner.[7] In response to a question from the Court, the Commission stated that 'Article 85(3) does not permit the Commission to conduct a cultural policy.'[8] This was discussed by the Advocate-General who concluded that 'there can therefore never be any question of the absolute priority of cultural over economic considerations' in the application of Article 85(3).[9] Despite the opportunity provided by the case, the Court made no comments on the relationship between cultural policy and the rules on competition. This reluctance is also evident in the second case, *Leclerc v Au Blé Vert*, concerning the application of Article 30, although Article 85 was also discussed.[10] This was a preliminary reference concerning the compatibility of a French law on Book Prices with various provisions of Community law.

The relevant law required all publishers or importers of books to fix the retail price for books which they publish or import. Retailers were required to charge an effective price for sales to the public of between 95 and 100 per cent of that price. Failure to do so would result in civil or criminal proceedings against the offending party. Civil proceedings were brought against Leclerc, a chain of low-price retail outlets, for selling books at prices below those fixed by law. Leclerc argued that Article 85 had been infringed, whereas the French Government considered the relevant provision to be Article 30.[11] The Court considered that the French law did not detract from the effectiveness of Article 85.[12] It continued by pointing out that, as Community laws stood, there was nothing to preclude the Member States from enacting legislation of the type at issue, 'provided that such legislation is consonant with other specific Treaty provision, in particular those concerning the free movement of goods'.[13] After analysing the law, the Court concluded that as it led to differences in treatment between domestic and imported books Article 30 had been breached. The French Government's attempt to justify the measure by pleading the mandatory requirement of consumer protection, referring in particular to the need to protect the role of books as a cultural media, was rejected. The Court specifically stated:

> As far as that point is concerned, it must be pointed out that national legislation which requires traders to abide by specific retail prices and discourages the marketing of imported products can be justified solely on the grounds set out in Article 36 of the Treaty. Since it derogates from a fundamental rule of the Treaty, Article 36 must be interpreted strictly and cannot be extended to cover objectives not expressly enumerated therein. Neither the safeguarding of consumers' interests nor the protection of creativity and cultural diversity in the realm of publishing is mentioned in Article 36. It follows that the justification put forward by the French government cannot be accepted.[14]

The Court avoided the opportunity of ruling on the question of whether cultural issues could come within the scope of the mandatory requirements, despite the fact that the justifications relied on by the French government related to these requirements.[15] The response of the Court was couched in the language of Article 36. Commenting on these two cases, the Commission concluded:

> The Commission has no intention of intervening in the cultural policy choices made by Member States, always provided of course that these choices do not result in measures being taken which conflict with Community law.[16]

The question to be addressed is therefore: how can a Member State exercise cultural policy choices which do not conflict with Community law? In the area of the free movement of goods, the obvious answer is that the Member State can seek to utilize the exceptions provided for in Article 36 of the Treaty through the protection offered by that article to national treasures or seek to extend the scope of the mandatory requirements to include cultural issues. As an alternative, it would be open to the Member States to seek Community legislation in the broad area of defending culture, for example in the context of the single market programme. The remainder of this chapter addresses the nature of these choices.

The judicial approach

Article 36, *inter alia*, allows for prohibitions or restrictions to the free movement of goods 'justified on grounds of the protection of national treasures possessing artistic, historic or archaeological value'. How this provision is to be interpreted in relation to the protection of culture is an open question. The first Commission Communication on Culture in 1977 made no reference to the scope of the exception in Article 36, a regrettable omission as the Court had not yet ruled on the scope of this exception to Article 36.

The possibility to do so arose in the 1968 case of *Commission v Italy* concerning the legality of a duty imposed by the Italian authorities on the export of art treasures.[17] The Italian government argued that the products subject to the measures were not goods within the meaning of the Treaty and, in the alternative, that the exception in Article 36 covered the particular measure. The Court began by pointing out that although the Treaty did not define goods, except to the extent that it covered everything which was not an agricultural product, it must be that anything capable of monetary valuation and being the subject of commercial transaction would constitute a 'good' for the purposes of the Treaty.[18] Having dismissed the first argument of the Italian government, the Court did not proceed to examine the scope of Article 36. That provision was not at issue, according to the Court, as the measure at issue was a tax, and Article 36 could not be relied on to justify a tax.[19] An opportunity to pronounce on the scope of the exception in Article 36 was therefore lost and, again, this is regrettable as the exact scope of the exception is by no means clear.

For example, although the English version of the Treaty refers to 'treasures', other versions of the Treaty refer to 'artistic heritage'. What, if any, is the significance of this difference in wording? Indeed, what is included within the phrase 'treasures'? Is there a distinction between treasures which are the products of the particular civilization and culture of a Member State and treasures which are accumulated by the Member State from other countries? Are treasures of regional significance, for example Scottish or Catalan treasures, included within the reference to national treasures. These questions of interpretation of Article 36 are addressed and answered by Oliver. He submits that there are three questions which must be asked in approaching this issue:

– Is the object of such historical or archaeological value to *this* Member State or any of its regions such as to qualify as a 'treasure'? If so, then the Member State is entitled to retain it without more ado, regardless of whether it has any artistic value. If not, then;
– can the object be described as an artistic treasure? If not then it falls outside this limb of Article 36. If it is an artistic treasure then;
– is it a product of the civilisation or culture of this Member State? If so, then it may be retained under this part of Article 36. If not, then one has to decide the point of principle whether the Member State concerned may retain the object nevertheless.[20]

This approach appears straightforward. It does vest considerable discretion in the Member States and, in the absence of harmonization, it still leaves a 'point of principle' to be resolved. It would have been helpful had the Commission chosen in the 1977 Communication to discuss the possible scope of the exception in Article 36 and to offer guidance on its interpretation and the point of principle recognized by Oliver. By 1982 and the second Commission Communication in this area, the Commission had at least recognized that there may be justified restrictions on the free movement of goods under this limb of Article 36 of the Treaty. However, they did not seek to address the question of how this provision could be used to protect culture.[21]

Although a considerable discretion is vested with the Member States under this limb of Article 36, there are limits on the exercise of that discretion. The European Court has recognized that as an exception to the principle of the free movement of goods, Article 36 must be interpreted strictly. The Member States must abide by the principles elaborated by the Court in their use of Article 36, whilst they are free to determine the scope of the phrase 'national treasures' and the measures to protect such treasures, these measures should be reasonable and proportionate to the goal to be achieved. It is for the Court to decide whether the measures exceed the bounds of reasonableness and/or proportionality so that the reconciliation of the protection of national treasures is not effected at an unjustifiable cost to the free movement of goods. The Court has considerable experience in this area deriving not only from its jurisprudence on Article 36 but also that relating to the mandatory requirements.

The concept of the mandatory requirements emerged from the decision of the Court in the *Cassis de Dijon* case.[22] The case concerned a requirement under German legislation that fruit liqueurs, like the *Cassis* which was to be imported in this case, could only be marketed if they contained a minimum alcohol content of 25 per cent. *Cassis de Dijon* has a minimum alcohol content between 15 and 20 per cent. The Court concluded:

> Obstacles to movement within the Community resulting from disparities between the national laws relating to the marketing of the products in question must be accepted in so far as those provisions may be recognised as being necessary in order to satisfy the mandatory requirements relating in particular to the effectiveness of fiscal supervision, the protection of public health, the fairness of commercial transactions and the defence of the consumer.[23]

The significance of this conclusion was that the Court made no reference to the issue of discrimination which had informed most of its previous decisions on the interpretation of Article 30. The issue of discrimination would, however, remain relevant to the interpretation of Article 34.[24] As a consequence of the decision, the non-discriminatory trade measures of the Member States will breach Article 30 unless they are necessary, and no more than is necessary, to satisfy one or more of the mandatory requirements.

Furthermore, the Court made clear that the category of mandatory require-ments was not limited to the four it had mentioned.

This case marked a major turning point in the approach of the Court to the interpretation of Article 30 and in the approach of the Commission to the issue of harmonization. From the point of view of the free movement of cultural goods, the case raises the question whether or not the protection of culture could be classified as a mandatory requirement and so be used to restrict the free movement of such goods. The 1982 Commission Commun-ication did not address this point and the Court, when presented with the question, in the case of *Cinéthèque*, also failed to give a clear answer.[25] At issue in this case was a 1982 French law on audio-visual communication which provided that no film shown in the cinema could be simultaneously sold or hired as a video until twelve months after its release in the cinemas. The question facing the Court was whether this amounted to a breach of Articles 30, 34, 36 and 59 of the Treaty. According to the Court, no issue was raised under Article 59, so the problems relating to the law fell within Articles 30–36. The Court concluded that the rule was prima facie within the scope of Article 30; it continued that it was 'lawful if there was objective justification of which it is acceptable with regard to Community law and provided also that the method of attaining the objective was proportion-ate'.[26]

What is significant about the decision is that it is unclear from the case which ground of justification was chosen by the Court to exclude this law from the scope of Article 30 (or Article 34). It could be argued that the Court was keen to promote and preserve 'cultural diversity' within the Com-munity.[27] This approach is helpful as it would also accommodate the Sunday Trading cases where the Court confirmed that national rules governing the opening hours of retail premises reflected:

> certain political and economic choices in so far as their purpose is to ensure that working and non-working hours are so arranged as to accord with national or regional socio-cultural characteristics and that, in the present state of Community law, is a matter for the Member States.[28]

The subsequent decision of the Court in *Keck and Mithouard*[29] casts great doubt on the continuing applicability of Article 30 to national rules govern-ing Sunday Trading.[30] This seems to be confirmed by the Court in its latest pronouncement on the matter in the case of *Semenaro*.[31] The decision in *Cinéthèque* remains to be explained; cultural diversity may still be an accept-able justification.

Advocate-General Slynn in delivering his opinion in *Cinéthèque* referred to the argument advanced by the Commission that, as films were part of contemporary culture, it was legitimate to adopt restrictions on the free movement of goods which would override the restrictions contained in Article 30. The justification for this was to preserve or support cultural activities. Advocate-General Slynn considered this to be a very broad principle, even assuming that cultural objectives could constitute one of the mandatory requirements. Having already decided on other grounds that the law at issue fell outside Article 30, Advocate-General Slynn concluded:

> If, therefore, I had not come to the conclusion that this law fell outside Article 30 for the first reason given, I would accept that it is capable of doing so if the measure adopted can be shown to be justified as being necessary for the maintenance of the film industry and the supply of films to the consumer. That objective is a legitimate one.[32]

Does this mean that the protection of culture is too broad a concept to be a mandatory requirement? After all the Advocate-General rejected the Commission's approach as being too broad and formulated the principle in terms of the 'maintenance of the film industry and the supply of films to the consumer'. Even this principle gives rise to questions. Which film industry is being supported – national, European or the entire film industry? Had the Court chosen to specify the ground of justification, and so deal with the issue of culture clearly, then some indication would have been forthcoming.[33]

The question which has to be asked is why did the Court not rule on this matter. Oliver offers the following explanation for the dilemma facing the Court:

> The problem is most delicate. On the one hand, the protection of culture is obviously a worthy objective which should have its place in Community law and which is by no means covered by the exception in Article 36 relating to 'the protection of national treasures having artistic, historic or archaeological value'. On the other hand, to bring whole sectors of commercial activity virtually outside Article 30 on the grounds that they involve the production of 'cultural' goods would be unthinkable. Moreover, 'culture' is notoriously difficult to define.[34]

The Court has consistently managed to avoid having to decide the question of the inclusion of the protection of culture within the mandatory requirements.[35] It is to be regretted that the Commission did not take the opportunity provided by the 1982 Communication, and indeed later Communications, to offer its opinion on this question. A partial solution to the problem of the interpretation of Article 36 would, however, be provided in the context of the implementation of the harmonization programme necessary to complete the internal market.

The legislative approach

In its 1987 Communication on Culture, *A Fresh Boost for Culture in the European Community*, the Commission proposed five aspects to the framework programme to cover the period up to 1992. Of these five, the creation of a European Cultural Area, was closely linked to the completion of the internal market programme. The Commission stated:

> The completion of the internal market implies – at a cultural level – the realisation of four major objectives, i.e.:
> (i) the free movement of cultural goods and services;
> (ii) better living and working conditions for those involved in cultural activities;
> (iii) the creation of new jobs in the cultural sector in association with regional development, notably in rural areas, tourism and technology;
> (iv) the emergence of a cultural industry which is competitive within the Community and in the world at large.[36]

The Council and the Ministers with responsibility for Cultural Affairs meeting within the Council highlighted only four priority areas, however it was noted that these four areas were 'without prejudice to actions which they consider desirable in other areas'.[37] The most surprising omission from the list of priorities was action to complete the cultural aspects of the internal market. The Commission, undeterred by this reverse, advocated a range of measures which they considered necessary to promote the free movement of goods, services and people in the cultural sector in the run-up to 1992.[38] Chief among these measures was the presentation of proposals for the definition of criteria which the Member States would use to identify 'national treasures' to which Article 36 of the Treaty could be applied and the implementation of procedures to guarantee the protection of national treasures. These proposals were duly submitted.[39]

The proposals were discussed at the November 1990 meeting of the Council and the Ministers with responsibility for Cultural Affairs. It was noted that with the abolition of controls on the free movement of goods at the end of 1992, trade in cultural goods would increase. The only protection would be afforded by Article 36 and the Member States agreed to a further examination of the impact of 1992 on the cultural aspect of this provision.[40] The end result of this further examination, and the Commission proposals, was the enactment of two measures related to the protection of national treasures. The first of these measures to be passed was Regulation 3911/92 on the export of cultural goods.[41] The primary purpose of the Regulation is to ensure that exports of cultural goods are subject to uniform controls at the external borders of the Community.

The means chosen to achieve this purpose is the requirement that all exports of cultural goods outside the customs territory of the Community be subject to the presentation of an export licence.[42] The licence, which is valid throughout the Community, is issued by a competent authority in the Member State in which the cultural object is located. Article 2(3) states:

> an export licence shall be issued at the request of the person concerned by the competent authorities of the member state on whose territory the cultural article in question was lawfully and definitively located on 1 January 1993, or, thereafter, on whose territory it is located following lawful and definitive export from another member state.

By necessary implication, this excludes stolen and illegally removed cultural objects. The licence may be refused according to Article 2(2) 'where the cultural goods in question are covered by legislation protecting national treasures of artistic, historical or archaeological value in the Member State concerned'.

The Annex to the Regulation lists the categories of cultural objects which are covered by the Regulation. The objects covered range from certain archaeological objects more than 100 years old to pictures and paintings to books more than 100 years old. Financial thresholds for certain categories of cultural goods are also set out in the Annex; for example, the financial threshold for books more than 100 years old is 50,000 ECU.[43] For those cultural objects not listed in the Annex or, if listed, valued below the financial thresholds, the national laws of the exporting country will continue to apply. The Annex does constitute an agreed list of common categories of cultural objects and indicates the extent to which the Member States are currently prepared to co-operate with each other in the protection of their respective cultures. However, the question of the interpretation of national treasure in the context of Article 36 remains unresolved by this Regulation.

The Regulation, according to Article 11, will enter into force three days after the publication of the Directive on the return of cultural objects unlawfully removed from the territory of a Member State. This Directive, Directive 93/7, was enacted in March 1993.[44] The Preamble recognizes:

> Whereas, under the terms and within the limits of Article 36 of the Treaty, Member States will after 1992 retain the right to define their national treasures and to take the necessary measures to protect them; whereas they will, on the other hand, no longer be able to apply checks or formalities at the Community's internal frontiers to ensure the effectiveness of those measures.

The Directive complements Regulation 3911/92 as it establishes a system to protect cultural objects within the single market. A cultural object is defined under Article 1(1) as falling within one of four categories. First, it may be an object which has been classified as possessing artistic, historic or archaeological value under national legislation. The next two categories concern objects which form an integral part of a public collection or the inventories of ecclesiastical institutions. For objects which do not fall within these latter categories, the Annex to the Directive lists further categories.[45] The approach of the Directive to the definition of 'cultural object' is therefore different from that of the Regulation to the definition of 'cultural goods'.

For objects unlawfully removed from the territory of a Member State, the Directive establishes a system guaranteeing Member States the return of national treasures. The Directive promotes co-operation between the competent authorities of the Member States and, as such, can be seen as a first step towards the mutual recognition of national laws on the protection of cultural objects. Under Article 5 of the Directive, the Member State from which the goods have been removed may initiate proceedings before a court in the Member State in which the goods are now located. Such proceedings must be brought not more than five years after the Member State from which the goods have been unlawfully removed becomes aware of their location. Article 8 makes clear that the proceedings may not be brought after a period of thirty years has elapsed since the unlawful removal of the cultural object. In the event that the court orders the return of the cultural object, the acquirer of the object will be entitled to 'fair compensation' having proved the exercise of all reasonable care at the time of the acquisition of that object.[46] Questions of ownership after the return of the object, according to Article 12 of the Directive, are to be governed by the law of the requesting Member State.

In a commentary on this legislation, von Plehwe recognizes that some degree of harmonization has been achieved and that this will promote some measure of uniformity within the Community.[47] He concludes that the legislation

> is in the nature of a compromise and symptomatic of the delicate balance of interests in an evolving 'principled market' that the solution adopted leaves numerous questions unanswered both in the application of the secondary legislation and in the area of national treasures not covered by the Regulation and the Directive.[48]

It is hoped that in the review of the effectiveness of Regulation by the Council every three years, the first of which took place in 1996, measures will be proposed by the Commission which will resolve some of the 'unanswered questions'.[49] The alternative avenue for resolving these questions, references on interpretation to the European Court, is not a promising option given the past reluctance of the Court.

Conclusions

The 1992 Commission Communication on culture gives some indication as to the future scope of Community action in the field of culture. It points out that:

> The Community is on the threshold of a new era in which it will be able to grow beyond its purely economic dimension and enjoy unprecedented opportunity for cultural co-operation and support ... thought should therefore be given to the future thrust of cultural action in this new environment.[50]

The Maastricht provisions echo this notion of the Community growing beyond its purely economic dimension and, indeed, they reflect the gradual extension of Community involvement in this area. Whether or not this gradual extension will involve a further diminution of the powers of the Member States to promote and protect their national cultures remains an open question.

In the first report on the consideration of cultural aspects in European Community action, the Commission states:

> [The] European Union, the principal elements of which have been linked historically to economic and commercial activities, is thus required to deepen, over a wider basis, which is likely to increase citizens' involvement and reinforce the sense of belonging to the European Union, whilst respecting the diversity of the national and regional traditions and cultures involved. In this respect, cultural action has a major role to play.[51]

Whilst the scope for Community action under Article 128(1) is limited to the areas mentioned, the possibility of action under paragraph (4) is, even when respecting the principle of subsidiarity, considerably larger. Returning to the problem outlined at the beginning of this chapter, it will not be easy to reconcile economic and cultural objectives given the predominance of economic objectives throughout the history of Community integration. However, such a reconciliation must be effected if the diversity of 'national and regional traditions and cultures' is to be respected. National cultures imply the existence of frontiers, the protection of such frontiers imply the acceptance of a limited degree of economic protection so as to promote the primary objective of the protection of cultures.

In the area of the free movement of goods, the two differing approaches are evident. The legislative approach necessitated by the creation of the internal market has been of a limited nature. However, a start has been made on the process of harmonization of the various laws of the Member States. The review of this legislation may lead to an extension of the scope of the Annexes to both the Regulation and the Directive, thus enlarging the scope of the co-operation between the Member States. By the very nature of culture, there will always be areas which will fall outside the scope of the legislation. In these areas and in the areas covered by the legislation, it will remain the responsibility of the Court to ensure that the Member States do not exceed the scope of their discretion in the protection and promotion of their national cultures.

In the jurisprudence of the Court the goal of promoting the integrity of the internal market remains paramount, although there are justifications for the protection of national markets which have been found to be acceptable. The Court's jurisprudence has ensured that protectionist measures within the Member States affecting the free movement of goods will continue to be eliminated. Justifications having a secondary effect of protectionism may be found to be acceptable. In relation to culture, the decisions in *Cinéthèque* and the Sunday Trading cases point to the limited acceptance by the Court of the need to allow for cultural diversity. However, there are limits to this cultural diversity, as the recent decision of the Court in *Piageme* makes clear.[52] At issue in this case was the language to be used for the sale of certain mineral waters imported from France and Germany in the Flemish-speaking area of Belgium. A 1986 Belgian Royal decree provided that the labelling should be in Dutch and the defendants contended that this legislation was contrary to, *inter alia*, Article 30 of the Treaty. The Court ruled that the legislation at issue, by making the use of a specific language compulsory, was contrary to Community law. It was a matter for the national courts to decide whether the language used on the particular product, other than the language of the region, was easily understood by consumers in that region. For the Court, the introduction of Article 128 by the Maastricht Treaty could not authorize a Member State to substitute more stringent rules that those laid down by existing Community law.

The introduction of Article 128 seems unlikely to change the thrust of the jurisprudence of the Court in this area. In the famous *Bosman* case, the Court rejected arguments based on the similarity between sport and culture by stating that:

> the question submitted by the national court does not relate to the conditions under which Community powers of limited extent, such as those based on Article 128(1), may be exercised but on the scope of the freedom of movement of workers guaranteed by Article 48, which is a fundamental freedom in the Community system.[53]

A similar interpretation seems likely to emerge in that most fundamental of all Community freedoms, the free movement of goods. In the light of its past jurisprudence, the Court will be more willing to support the right of the individual to trade freely within the Community rather than the interests of groups of individuals in the maintenance of cultural diversity within the

Union. For the Court to rule otherwise would require it to redefine the scope of Article 30, so upsetting a long line of consistent jurisprudence. Although the Court in *Keck* altered its jurisprudence on Article 30 by renouncing some level of control over national regulatory measures, to admit that the protection of cultural diversity can allow for the non-application of Article 30 would allow too great a power to the Member States in an area of exclusive Community competence. For this reason the protection of cultures should be actively pursued by the Member States through the legislative approach rather than leaving the matter to individuals and the Court of Justice. Such an approach is consistent with Article 128 which vests the primary responsibility for the protection of cultures not with the Community but with the Member States.

Notes

1. *Financial Times*, 16 February 1996, p. 3.
2. Joined cases 43/82 and 63/82 [1984] ECR 19.
3. *Ibid.*, pp. 33–4.
4. *Ibid.*, p. 55. The Court stated that the agreement 'involved a restriction on competition within the common market by reason of both the exclusive dealing system and the collective resale price maintenance for which it makes provision'.
5. *Ibid.*, p. 62. The Court concluded: 'To submit the production of and trade in books to rules whose sole purpose is to ensure freedom of trade between Member States in normal conditions of competition cannot be regarded as restricting freedom of publication which, it is not contested, remains entire [*sic*] at the level of both publishers and distributors.'
6. *Ibid.*, p. 63.
7. See OJ 1982 L54/36, para. 60. The Advocate-General noted that in response to questions asked by the Court, the Commission claimed: 'In appreciating an agreement in the light of Article 85(3) it neither can nor may attribute decisive significance to certain culturally desirable consequences of a given form of distribution without more ado.' (*Ibid.*, p. 89.)
8. *Supra*, n. 2, p. 48.
9. *Ibid.*, p. 89.
10. Case 229/83 [1985] ECR 1.
11. *Ibid.*, pp. 19–20.
12. *Ibid.*, p. 31. The Court noted: 'Whilst it is true that the rules on competition are concerned with the conduct of undertakings and not with national legislation, Member States are none the less obliged under the second paragraph of Article 5 of the Treaty not to detract, by means of national legislation, from the full and uniform application of Community law or from the effectiveness of its implementing measures; nor may they introduce or maintain in force measures, even of a legislative nature, which may render ineffective the competition rules applicable to undertakings.'
13. *Ibid.*, p. 33.

14. *Ibid.*, p. 35.
15. See discussion below in relation to the mandatory requirements, especially text accompanying notes 22–35.
16. COM (85) 681 *The European Dimension with Regard to Books*, p. 6.
17. Case 7/68 [1968] ECR 423.
18. *Ibid.*, p. 428.
19. *Ibid.*, p. 430.
20. Oliver, P. (1988) *The Free Movements of Goods* (Second edition), London: ELC, p. 205. These criteria bear some resemblance to the Waverly criteria, a result of the recommendations of the Waverly Committee, which are used by the United Kingdom and which ask: (i) is the object so closely connected with history and national life that its departure would be a misfortune? (ii) Is it of outstanding aesthetic importance? (iii) Is it of outstanding significance for the study of some branch of art, learning or history?
21. Instead, the Commission argued that the traditional concept of 'national heritage' inherent in this exception should be expanded to become the new concept of 'Community heritage'. EC Bulletin Supp. 6/82, *Stronger Community Action in the Cultural Sector*, p. 8.
22. Case 120/78, *Rewe Zentrale v Bundesmonopolverwaltung für Branntwein* [1979] ECR 649.
23. *Ibid.*, p. 662.
24. Case 237/82 *Kaas* [1984] ECR 483.
25. Cases 60 and 61/84 [1985] ECR 2605.
26. *Ibid.*, p. 2628.
27. This justification is the one adopted by the Commission. See COM (96) 160, *First Report on the Consideration of Cultural Aspects in European Community Action*, p. 25.
28. Case C-145/88, *Torfaen B C v B&Q plc* [1989] ECR 3851, p. 3889. See also case C-312/89, *Conforma* [1991] ECR I-997, case C-322, *Marchandaise* [1991] ECR I-1027, case C-169/91, *Stoke-on-Trent v B&Q plc* [1992] ECR I-6635.
29. Cases C-267 and 268/91 [1993] ECR I-6097. This decision has attracted considerable comment; see D. Chalmers (1994) 'Repackaging the Internal Market: The Ramifications of the *Keck* Judgment', in *ELRev*, 385; N. Reich (1994) 'The "November Revolution" of the European Court of Justice: *Keck, Meng*, and *Audi* Revisited', in *CMLRev*, 31, 845; and M. Ross (1996) '*Keck*: Grasping the Wrong Nettle', in A. Caiger and D. Floudas (eds) *Onwards: Lowering the Barriers Further*, New York: Wiley, pp. 45–62.
30. See M. Jarvis (1995) 'The Sunday Trading Episode: In Defence of Euro-Defence', in *ICLO*, **44**, 451.
31. Joined Cases C-418/93 to C-421/93, C-454/93, C-9/94 to C-11/94 and C-15/94, C-23/94 and C-332/94, *Semenaro Casa Uno Srl e.a. v Sindaco del communi di Erbusco e.a.*, Decision of 20 June 1996 (unreported).
32. *Supra*, n. 25, p. 2613. As for the ground for deciding that the law did not fall within Article 30, Advocate-General Slynn commented: '. . . in an area in which there are no Community standards, where a national measure is not specifically directed at imports, does not make it any more difficult for an importer to sell his products than it is for a domestic producer, and gives no protection to domestic

producers, then in my view, the measure does not fall within Article 30 even if it does lead to a restriction or reduction of imports'.

33. In Case C-17/92, *Federaçion de Distribuidores Cinematograficos v Estado Espanol* (Decision of 4 May 1993, not yet reported), a request from the Spanish Supreme Court for a preliminary ruling on the compatibility with Community law of national rules subjecting the grant of licences to dub films for distribution within Spain. The Court ruled that the matter fell within Article 59 and could only be justified under the exceptions provided in Article 56. The Court refused to accept the Spanish Government's argument that the measure pursued the cultural aim of the protection of the domestic film industry by pointing out that cultural policy was not one of the justifications listed in Article 56.

34. *Supra*, n. 20, p. 221.

35. See also Case 229/83, *Leclerc v Au Blé Vert* [1985] ECR 1, discussed above; see text accompanying notes 10–14. However, the Court has stated in the context of the provision of services that 'The general interest in consumer protection and in the conservation of the national historic and artistic heritage can constitute an overriding reason justifying a restriction on the freedom to provide services.' Case C-180/89, *Commission v Italy* [1991] ECR I-709, at I-723. The formulation of this exception in Case C-154/89, *Commission v France* [1991] ECR I-659, was slightly different: 'The general interest in the proper appreciation of places and things of historical interest and the widest possible dissemination of knowledge of the artistic and cultural heritage can constitute an overriding reason justifying a restriction on the freedom to provide services', p. I-687. Similarly, in Case 198/89, *Commission v Greece* [1991] ECR I-727, the Court again reformulated the exception by stating: 'The general interest in the proper appreciation of the artistic and archaeological heritage of a country and in consumer protection can constitute an overriding reason justifying a restriction on the freedom to provide services.' In all three cases, the restriction imposed was held to go beyond what was necessary to protect the general interest (p. I-741).

36. EC Bulletin Supp., 4/87, p. 10.

37. OJ 1988 C-197/2, para. 2.

38. *Supra*, n. 36, p. 10.

39. COM (89) 594, *Commission Communication on the Protection of National Treasures Possessing Artistic, Historic or Archaeological Value: Needs Arising from the Abolition of Frontiers in 1992*. On the same topic, see COM (91) 447. In this *Communication*, the Commission stated: 'It would be inconceivable to apply unrestrictedly the logic of the internal market and the principle of the free movement of goods in respect of objects that constitute national treasures: account must be taken of the special nature of cultural items, which cannot be treated as mere goods. The fact remains, however, that completion of the internal market could be rendered difficult by the implementation of national measures aimed at protecting national treasures if those measures infringed Article 36' (p. 2).

40. EC Bulletin 11/1990, point 1.3.187.

41. OJ 1992 L395/1.

42. *Ibid.*, Article 2(1).

43. According to Article 10(5) of the Regulation, these thresholds will be adapted to economic and monetary developments as necessary.

44. OJ 1993 L74/1.
45. The categories are roughly comparable to those in the Annex to Regulation 3911/92.
46. *Supra*, n. 44, Article 10.
47. Some degree of harmonization and uniformity may also be promoted as a result of Council Decision 95/468 (OJ 1995 L269/1) establishing the IDA programme. Under this programme, which extends beyond the strict confines of cultural objects, administrative co-operation is promoted between the customs administrations of the Member States through the facilitation of exchanges of information.
48. 'European Union and the Free Movement of Goods', in *ELRev*, **20**, 430 (1995), pp. 449–50.
49. The Commission has already proposed amendments to the Annexes of Regulation 3911/92 and Directive 93/7 ahead of the triennial review as a result of the fact that language differences and different cultural traditions of the Member States have resulted in problems with the classification of 'water colours, pastels and gouaches'. The proposal involves the creation of a new category to cover this category of cultural goods. COM (95) 479. As for the results of the review, they were limited to an extension of the scope of the Annex to Directive 93/7 (OJ 1997 L 60/59).
50. COM (92) 149, *New Prospects for Community Cultural Action*, p. 1.
51. *Supra*, n. 27, p. 1.
52. Case C-85/94, *Piageme v Peeters* [1995] ECR I-2955.
53. Case C-415/93 *Bosman* [1995] ECR I-4921.

Chapter 14

Languages and the European Union

JOHN A. USHER

Introduction

Law and problems of language sometimes seem to be synonymous in the European Community context. The Community of fifteen Member States uses eleven working languages: French, German, Italian, Dutch, English, Danish, Greek, Spanish, Portuguese, Finnish and Swedish. While this list does not include every official language of every Member State, it does at least include *one* official language from each Member State. With the exception of the 1951 European Coal and Steel Treaty, of which only the French text is authentic,[1] all Community legislation is equally authentic in all eleven languages. From time to time there have been complaints about the cost of this process, but in a system of law which envisages common policies applying throughout the whole Community and which creates legal instruments[2] which are binding as law in every Member State simply because they have been adopted by the Council and Parliament, Council or Commission, as the case may be, without any need for re-enactment by the national authorities, it may be suggested that it is inconceivable that such legislation should not be available in a language spoken and understood by each citizen to which it applies.

This in turn gives rise to the practical problem that if the Community legislation is to apply in the same way and at the same time in every Member State, it cannot be brought into force, and indeed will not be printed in the Official Journal, until every language version is available. This creates immense pressures for the translation services of the Council and Commission, pressures which have led the Commission to pioneer computerized translation systems. These systems have not been without their problems, and if what one reads in the *Guardian* is to be believed, it would appear that at the experimental stage they achieved such feats as turning the phrase 'les agriculteurs vis-à-vis de la politique agricole commune' into 'farmers live to screw the common agricultural policy'. One is tempted to say that many a true word is spoken by a mis-programmed computer.

It may be added that successive Acts of Accession have been drafted so as to ensure that translations of existing Community legislation into the languages of new Member States are just as authentic as the original versions.[3] In some cases one is tempted to think that the translation has been used to resolve problems arising from the original text, as in the English version of Article 58 of the EC Treaty. This provision extends the freedom of establishment conferred by the Treaty on Community nationals, that is, the freedom to set up business in another Member State or to create a subsidiary, branch or agency there, to other forms of business enterprise. The original versions used a single word such as 'société' or 'Gesellschaft', leaving commentators to speculate whether it was only those types of undertaking which enjoyed legal personality which could take advantage of freedom of establishment. However, the English version refers to 'companies or firms', a form of words apt to include partnerships, which under English law do not have legal personality, and presumably the other versions must be interpreted the same way.

On the other hand, such translations may create unintended problems. A particularly striking example was Case 55/74 *Unkel v Hauptzollamt Hamburg-Jonas*.[4] This was a reference for a preliminary ruling from a German court. A question was asked, amongst other things, about a particular provision of a regulation[5] governing export refunds, the English text of which read that 'the time limit for claiming payment of the refund shall be six months'. The German text, however, giving a literal translation of it, said that 'the documents in support of the claim must be presented within six months'. The question effectively asked what these documents were, a question nobody would have thought of asking on the basis of the English text. Indeed, the translator of the English version of the judgment had to insert an embarrassing note to the effect that the English text of the regulation made no reference to the 'documents'. Since the tenor of all the other language versions was similar to that of the German version, the Court was in fact able to answer the question put to it. It might be added that the re-enactment of that provision in a later regulation had an English text which was parallel to the other language versions.[6]

Translation problems have even led the European Court not to apply the principle that ignorance of the law is no defence. In a reference from a Greek court to the European Court shortly after Greek accession,[7] the fact that a Greek language text of the relevant Community legislation was not available to the public, that local civil servants had not received instructions on the matter, and that the head of service was absent at the relevant time were held to constitute 'special reasons' in the context of legislation allowing the remission of import and export duties in situations arising from special circumstances in which no negligence or deception could be attributed to the

person concerned. In particular, it was expressly stated that small undertakings far from Athens could not, in that situation, be expected to know the
Community rules for themselves.

However, while all this may be true of legislation as such, chinks are
beginning to appear at other levels of Community activity. Thus, while the
European Court and the European Court of First Instance allow all the
working languages mentioned above (and Irish) to be used as languages of
procedure in cases before them, it is no longer the case that all their
judgments are automatically published in all those languages. While it may
seem a relatively unimportant exception, judgments in staff cases (that is,
disputes between Community officials and their employing institution) are
only published in full in the language in which they were heard (which is
technically the authentic text). Unless the case is regarded as being of
particular importance or interest, however, only a summary is published in
the other languages.[8] Perhaps of wider concern is the fact that there are now
entities created under EC law which do not allow all the languages mentioned above to be used for all purposes. The example may be given of the
Office for Harmonisation in the Internal Market (trade marks and designs)
established under the EC Regulation on the Community trade mark.[9] While
applications for a Community trade mark may be filed in any of the official
languages, the 'languages of the Office' are defined as English, French,
German, Italian and Spanish. An applicant is required to indicate a second
language which is one of those languages. If the application was in another
language, the Office may then write to the applicant in that second language,
and the applicant is required to accept it in opposition, revocation or
invalidity proceedings. This was challenged by a Dutch lawyer before the
Court of First Instance[10] and on appeal to the European Court,[11] but the first
instance action was held inadmissible, and the appeal was held manifestly
unfounded, so the substantive issue has not yet been determined by the
courts.

Interpretation and integration

The need to ensure substantive uniformity despite the use of eleven languages is one reason why the EC Treaty gives the European Court
jurisdiction to interpret provisions of Community law at the request of any
national court or tribunal, a jurisdiction which has been used to develop the
concepts of direct effect (that is, private enforcement) and the primacy of
Community law, which have been of enormous importance in the development of legal integration.

A salutary example of what can happen in the absence of provision for
uniform interpretation may be found in 1956, when the EEC Treaty was

being negotiated. In January 1956, the French Cour de Cassation[12] had to interpret the phrase 'l'aval est réputé donné pour le tireur' in the Code de Commerce. Realizing that this phrase was the French implementation of a provision in the Geneva Convention on bills of exchange, the court looked at the other language versions. Normally in French law, the verb 'réputer' creates a simple presumption, and evidence to the contrary is admitted, but noticing that the German version used a verb which creates an irrebuttable presumption, so that no evidence to the contrary is admitted, it was held that in this context 'réputé' must create an irrebuttable presumption. In November 1956, the German implementation of the same provision fell to be considered by the Bundesgerichtshof.[13] Out of the best of internationalist motives, it looked also at the French version, and in the light of that interpreted a German word which would normally create an irrebuttable presumption as creating only a simple presumption. Hence, for the best of reasons, the same provision of the convention was interpreted differently in France and Germany; it may be wondered to what extent this influenced those who drafted the EEC Treaty.

However, the situation has often arisen where a national court has not referred a matter to the European Court because it has taken the view that the meaning of the Community provision is so obvious that no question as to its interpretation arises. For its part, the European Court has recognized that the correct application of Community law may be so obvious as to leave no reasonable doubt as to the manner in which the question raised is to be resolved, but it went on to add that the national court must be convinced that the matter would be equally obvious to the courts of the other Member States and to the European Court itself, bearing in mind in particular that Community legislation is drafted in several equally authentic languages, that Community law uses its own terminology and that legal concepts do not necessarily have the same meaning in Community law and in the laws of the Member States.[14] Few national courts could claim the linguistic expertise to be able to comply with that requirement, and the underlying message is clearly that novel points of Community law should be resolved by the European Court rather than by national courts. Indeed, it means that a national court cannot treat the interpretation of Community law in a purely national context.

It may be suggested that it is to a large extent language problems which have led the European Court to use a purposive rather than a literal approach to the interpretation of Community legislation. Where there are divergences between different language versions, it is highly unlikely that a satisfactory result can be obtained by just looking at the words alone, even where a limited range of languages may be at issue. The original version of the Common Customs Tariff was based upon the 1951 Customs

Co-operation Council Nomenclature Convention, of which the only authentic texts were French and English,[15] a fact of which the Court took account. The example might be taken of Joined Cases 824 and 825/79 *Folci v Italian Finance Administration*[16] on the interpretation of CCT Heading 07.04 B in the context of two Council regulations on generalized tariff preferences in favour of developing countries, allowing preferential treatment to '*whole* mushrooms, dried, dehydrated or evaporated, excluding cultivated mushrooms' under that heading. The question raised was whether 'whole' meant 'in one piece', the view of the customs authorities, or 'the entire mushroom, albeit in pieces', the view of the importer. In answering this, the Court looked to see whether the CCT heading itself recognized a distinction between intact vegetables and those in pieces. The importer had relied upon the Italian text, which was in turn based on the French version of heading 07.04 B:

> Légumes et plantes potagères desséchés déshydratés ou évaporés, même coupés en morceaux ou en tranches ou bien broyés ou pulvérisés, main non autrement préparés . . .

This, although it includes sliced and cut vegetables, does not expressly distinguish them from intact vegetables. However, the Court, having regard to the authentic languages of the CCC Nomenclature Convention, looked also at the English version: 'Dried, dehydrated or evaporated vegetables, whole, cut, sliced, broken or in powder but not further prepared.' It noted that here 'whole' was differentiated from 'cut' or 'sliced', and hence that a 'whole' vegetable could not be 'cut' or 'sliced'. To confirm this view, however, the Court looked also to the *purpose* of the Regulations, noting that they were intended to prevent the importation at preferential rates of cultivated mushrooms, and that it would be virtually impossible to distinguish wild and cultivated mushrooms if they were cut or sliced. Hence it concluded that 'whole' here must mean 'in one piece'.

The link between language problems and the 'purposive' or 'teleological' system of interpretation which, as is widely known, is much used by the European Court, has indeed been recognized by the Court itself. In case 61/72 *Mij. PPW International v Hoofdproduktschap voor Akkerbouwprodukten*,[17] where the obligations of national authorities with regard to the dispatch of 'advance fixing certificates' were at issue, the Court stated that:

> No argument can be drawn either from any linguistic divergences between the various language versions, or from the multiplicity of the verbs used in one or other of those versions, as the meaning of the provisions in question must be determined with respect to their objective.

Looking at the objectives of the system of advance fixing certificates, the Court noted that they were only issued on payment of a deposit by the trader, and that payment of, in this case, export refunds, depended on the presentation of a certificate by the trader. From this, it held that the national authorities had a duty to ensure that such certificates actually reached applicants for them, and that this obligation was not fulfilled by sending them by post when they failed to reach the addressee.

An example of this purposive interpretation arising from linguistic divergence involving the United Kingdom[18] arose from the fact that a group of British trawlermen wished to obtain some cod, for which they were not allowed to fish in Community waters, but which did happen to be available in the area of the Baltic over which Poland claimed exclusive fishing rights, and where British trawlers had no right to fish. At the same time Polish trawlermen were wanting herring and mackerel, which were not available in the Polish area of the Baltic, but which could be caught in Community waters – except that Polish boats had no right to fish there. A group of British trawlers therefore set off for the Baltic laden with herring and mackerel, and met the Polish trawlers off the Polish coast. The British trawlers cast their empty nets into the sea, and these nets were then taken over by the Polish boats, which trawled them but did not take them on board. After the trawl was completed, the Polish trawlers passed the ends of the nets to the British trawlers, and the cod was landed on to the British trawlers; in return for this, the British boats transferred the herring and mackerel to the Polish boats.

When the British trawlers returned to the United Kingdom, the customs authorities treated the cod as being British, and therefore not liable to pay common customs tariff duties. The reason for this was that under the EEC Regulation determining the origin of goods for customs purposes, fish are treated as wholly obtained in one country if they are 'taken from the sea' by vessels registered in that country and flying its flag. The British view essentially was that since the nets were actually pulled from the sea by the British trawlers, the fish were 'taken from the sea' by the British trawlers, whereas the Commission's argument was that fish were taken from the sea when the net closed round them, irrespective of when they were physically hauled out of the sea.

Faced with this dispute, the Court first looked at the texts of the Regulation, and noted that the French version used the phrase 'extraits de la mer', which appeared to support the British argument, whereas, for example, the German version used the word 'gefangen', meaning caught, which tended to support the Commission's argument. After referring also to the Greek, Italian and Dutch versions, but not, it would appear, the Danish(!), the Court concluded that a comparative examination of the various language versions did not enable a conclusion to be reached in favour of any of the arguments put forward, and so no legal consequences could be based on the terminology used. It therefore expressly turned to consider the purpose and general scheme of the Regulation determining the origin of goods for customs

purposes, and came to the conclusion that in the context of a fishing
operation carried out by a number of vessels registered in different countries,
the origin should in principle depend on the flag flown by the vessel which
performed the *essential part* of the operation of catching them. Faced with the
fundamental question of legal philosophy as to when a free object becomes
property, the Court took the view that the essential part of catching fish is
locating the fish and separating them from the sea by netting them, and that
simply hauling the nets out of the sea is not the essential part of the
operation. Whatever may be thought of the Court's analysis, this judgment
at least illustrates its general approach to such problems.

Languages and the free movement of persons

In a collection of essays essentially concerned with frontiers, particular
attention may be paid to the extent to which language problems have
affected the aim of the Community to become an area without internal
frontiers in which there is free movement of goods, persons, services and
capital. Leaving on one side the practical problem that personal rights of
movement may be of little utility to those who do not have the linguistic
ability to take full advantage of them, it is now clear that there are some
activities where a knowledge of the (or even a) local language may be
imposed as a precondition, thus effectively restricting the apparent Treaty
freedom of movement. Furthermore, the case law appears to indicate that
where there is a conflict between the protection of local culture and freedom
of movement under the EC Treaty, the protection of local culture may
prevail. The matter arose in Case C-379/87 *Groener v Minister for Education
and the City of Dublin Vocational Education Committee*,[19] which involved Irish
measures requiring full-time appointees to certain teaching posts to hold a
certificate of proficiency in the Irish language. Mrs Groener was a Dutch
national who had applied for a full-time post as an art teacher but failed an
Irish language test, with the result that the Minister refused to appoint her to
the post. While in principle the right of a national of one Member State to
seek employment in another Member State under the EC Treaty is subject
to local conditions of employment,[20] the EC implementing legislation[21] states
that such national provisions are not to apply where 'their exclusive or
principal aim or effect is to keep nationals of other Member States away from
the employment offered' unless they relate to 'linguistic knowledge required
by reason of the nature of the post to be filled'. On the evidence before it, the
European Court found that the teaching of art in public vocational education
schools in Ireland was conducted essentially, or indeed exclusively, in the
English language. One might therefore have expected the Court to conclude
that an Irish language requirement for an art teacher did not relate to

'linguistic knowledge required by reason of the nature of the post to be filled'. However, it took the view that this phrase went beyond simple practical necessity, stating that although Irish was not spoken by the whole Irish population, the policy followed for many years by Irish governments had been designed not only to maintain but also to *promote* (emphasis added) the use of Irish as a means of expressing national identity and culture. The Court then held that the Treaty did not prohibit the adoption of a policy for the protection and promotion of a language of a Member State which was both the national language and the first official language, provided that the requirements deriving from measures intended to implement such a policy were not disproportionate in relation to the aims pursued, and provided also that they did not bring about discrimination against nationals of other Member States.

From these criteria the Court concluded that given the importance of education for the implementation of such a policy, it was not unreasonable to require teachers to have some knowledge of the first national language, provided that the level of knowledge required was not disproportionate, that the power to grant exemptions was exercised in a non-discriminatory manner, that nationals of other Member States were not required to be present in the host State while learning the language and that nationals of other Member States had an opportunity to retake the examination if they applied for another post covered by the language requirement. In such circumstances, a measure such as the Irish language requirement could be regarded as relating to 'linguistic knowledge required by reason of the nature of the post to be filled'.

While the present chapter is not the place to comment on this judgment as an example of detailed judicial law-making, it does clearly show that national language requirements may be allowed to restrict a fundamental Treaty freedom not simply on grounds of practical necessity but also as a matter of cultural policy, and that the European Court was willing so to decide several years before the Maastricht amendments introduced the concept of cultural policy into the text of the EC Treaty.[22]

On the other hand, it is also clear that those who have taken advantage of the Treaty rights of free movement may enjoy options as to the use of language, at least in Member States with more than one official language, which are not available to home citizens whose situation does not fall within the Treaty rules. In this context, the European Court has found itself faced with national politico-linguistic problems, such as the Flemish/French division in Belgium, which was encountered in Case 55/77 *Maris v Rijksdienst voor Werknemerspensioenen*,[23] where the Community rules on social security for migrant workers were at issue. The claimant here was a Belgian national resident in France who had completed insurance periods in Belgium,

Germany and France. In the course of a dispute with a Belgian pension institution, she brought an action before the local court in Antwerp, the cause of the reference to the European Court being that she filed her application in French, whereas under Belgian legislation the language to be used in the Antwerp court was Dutch, and that court was required to declare of its own motion that any pleading drawn up in a language other than the official language of the court was null and void. Maris, however, invoked Article 84(4) of Council Regulation 1408/71 on social security for migrant workers, providing that

> The authorities, institutions and tribunals of one Member State may not reject claims or other documents submitted to them on the grounds that they are written in an official language of another Member State.

Recognizing the potential problems caused by the fact that the claimant here was herself Belgian, the European Court nonetheless stated that:

> Having regard to the large number of different individual situations to which freedom of movement for workers and their families may give rise, Article 84(4) does not for reasons of practicability draw any distinction based on the nationality of the persons concerned or on their residence as long as the purpose of the claims submitted or the documents produced is the implementation of the regulation in question.

It did, however, rather defensively emphasize that Article 84(4) only applied to claims submitted by persons covered by the Regulation, that is, workers who have moved between two or more Member States, and only in relation to the implementation of the Community rules on social security, the general procedure and resolution of other disputes involving workers remaining governed by domestic law. Although this was a fairly straightforward interpretation of Article 84(4) of Regulation 1408/71, it nonetheless apparently made the headlines in the Flemish press because as a result a French-speaking Belgian was able to use French before a Belgian court in a Flemish-speaking area.

Languages and the free movement of goods and services

Other tensions may be found in the area of consumer protection. While it has long been accepted that non-discriminatory consumer protection rules may be enforced against goods imported from other Member States, under the case law of the Court which recognizes that such measures may be justified to protect 'mandatory requirements',[24] it has usually been held that such protection can satisfactorily be achieved by labelling requirements rather than by prohibiting the entry of the goods in question. Thus Belgium could not require margarine to be sold in cubic blocks to distinguish it visually from butter sold in oblong blocks (effectively closing the Belgian market for margarine) if it was clearly labelled as margarine.[25] Many more examples could be given: German law requiring only Franconian wine to be sold in the 'Bocksbeutel' shape of bottle could not be used to prevent the import of Italian wine labelled as such which was normally and lawfully sold in that shape of bottle on its home market;[26] Italian law requiring pasta to be made from durum wheat could not prevent the importation of pasta containing soft wheat provided it was labelled as such (even though the Court recognized that Italian consumers would be unlikely to eat it),[27] and German law on the meat content of sausages could not be enforced against imported sausages where there was no threat to health and consumers could be informed of the contents by labelling.[28] Similarly, French legislation restricting the use of the word 'yoghurt' to a live product could not stop the sale of frozen yoghurt,[29] and rules on the minimum fat content of Edam cheese could not stop the sale as Edam of imported cheese with a lower fat content if it was so labelled.[30]

However, it now seems to be established in the case law of the European Court that, depending on the nature of the goods, labelling need not necessarily be in the official language of the area where the goods are sold, provided it is in a language which may easily be understood by consumers in that area – a concept which seems to raise more questions than it resolves. The concept is clearly set out in Community legislation in Council Directive 79/112[31] which requires various foodstuffs only to be labelled 'in a language easily understood by purchasers'. In Case C-369/89 *Piageme v Peeters*[32] which has recently been reaffirmed[33] it was held that, by virtue of this Directive, Belgium could not require French and German mineral water labelled in those languages to be labelled in Dutch when sold in the Flemish area of Belgium without allowing for the possibility of using another language easily understood by purchasers.

In fact, the actual terms of the Directive are at first sight restrictive rather than permissive. Article 14 states that:

> The Member States shall, however, ensure that the sale of foodstuffs within their own territories is prohibited if the particulars provided for . . . do not appear in a language easily understood by purchasers, unless other measures have been taken to ensure that the purchaser is informed. This provision shall not prevent such particulars from being indicated in various languages.

In the first *Piageme* case, the Court held however that this legislation precluded national law from requiring the exclusive use of a specific language for the labelling of foodstuffs, without allowing for the possibility of using another language easily understood by purchasers or of ensuring that the purchaser was informed by other measures. In the second case, the Court confirmed that the expression 'a language easily understood' was not equivalent to the official language of the Member State or the language of the region, and pointed out that there was other Community legislation which did expressly require labelling in the language or languages of the Member State where the product was placed on the market, such as Council Directive 92/27 on the labelling of medicinal products for human use.[34] Reference was also made to two of the provisions introduced into the EC Treaty by the Maastricht amendments, Article 128 on cultural policy, which requires the Community to respect national and regional cultural diversity, and Article 129(a) on consumer protection, requiring the Community to contribute to a high level of consumer protection. Despite its previous willingness to allow the Irish language requirement for teachers to be defended as a matter of cultural policy, the Court took the view that neither of these provisions authorized a Member State to substitute a more stringent rule for that laid down in the Directive. However, the Court did hold that it was for the national court to determine in each individual case whether the required information given in a language other than the language mainly used in the Member State or region concerned could be easily understood by consumers in that State or region; it was further suggested that relevant factors might include the possible similarity of words in different languages, the widespread knowledge amongst the population concerned of more than one language, or the existence of special circumstances such as a wide-ranging advertising campaign or widespread distribution of the product, 'provided that it can be established that the consumer is given sufficient information' – a rider which appears to beg the issue.

However, while the purchase of mineral water may not pose too many risks for the consumer, there are some more general issues which arise. Community and national law requirements to use a local language or a 'language easily understood' apply only to products which are actively marketed in the area in which that legislation applies. The idea of the internal market in the EC, on the other hand, involves not just the freedom for a trader in one Member State to sell goods in another Member State, but also the freedom for a purchaser resident in one Member State to seek out and purchase goods. Such a purchaser will receive no linguistic protection. Furthermore, similar freedoms extend to the provision of services within the EC,[35] even the delicate area of financial services.[36] Thus a policyholder may seek out insurance cover from an insurer in another Member State, and it has now been clearly held that a resident of one Member State has a Community law right to take out a loan, even a loan for house purchase, from a lender established in another Member State[37] – transactions which many consumers would have difficulty in understanding even in their own language.

Conclusion

It has been suggested that language problems have contributed to the development of purposive (and therefore, it may be submitted, by definition integrationist) interpretation of EC legislation by the European Court of Justice. On the other hand that same Court has been willing to allow certain national language requirements to be defended, even where they interfere with the economic freedoms laid down by the EC Treaty, to the extent that this can be justified as a matter of cultural policy. By way of contrast, local language requirements may not be enforceable as such in the context of the marketing of certain low-risk products, and consumers who seek goods and services in other Member States may receive no linguistic protection at all.

Dare it be suggested that the creation of a true single market necessarily implies a considerable broadening of linguistic capabilities?

Notes

1. Article 100.
2. Notably 'regulations' as defined under art. 189 of the EC Treaty.
3. In the case of English, see arts. 155 and 160 of the 1972 Act of Accession.
4. [1975] ECR 9.
5. Commission Regulation 1041/67/EEC (OJ 1967 314/9).
6. Commission Regulation (EEC) 2110/74 (OJ 1974 L220).
7. Case 160/84 *Oryzomyli Kavallas* [1986] ECR 1633.
8. See the explanatory note in [1994] ECR-SC.
9. Council Regulation 40/94 (OJ 1994 L11/1).
10. Case T-107/94 *Kik v Council and Commission* [1995] ECR II-1717.
11. Case C-270/95 P *Kik v Council and Commission* (28 March 1996).
12. Cass. Comm. 23 Jan. 1956 (JCPII 9666).
13. Bundesgerichtshof 15 Nov. 1956 (1962 NJW 745-6).
14. Case 283/81 *CILFIT v Italian Ministry of Health* [1982] ECR 3415.
15. Convention Article XVI.
16. [1980] ECR 3053.
17. [1973] ECR 301.
18. Case 100/84 *Commission v U.K.* [1985] ECR 1170.
19. [1989] ECR 3967.
20. EC Treaty art. 48(2), unless the rules at issue make freedom of movement impossible, Case C-415/93 *Union Royale Belge des Sociétés de Football Association v Bosman* [1995] ECR I-4921.
21. Art. 3(1) of Council Regulation 1612/68.
22. Art. 128.
23. [1977] ECR 3961.
24. Generally known as the 'Cassis de Dijon' doctrine after the subject-matter of Case 120/78 *Rewe v Bundesmonopolverwaltung für Branntwein* [1979] ECR 649.
25. Case 261/81 *Rau* [1982] ECR 3961.
26. Case 16/83 *Prantl* [1984] ECR 1299.

27. Case 90/86 *Zoni* [1988] ECR 4285.
28. Case 274/87 *Commission v Germany* [1989] ECR 229.
29. Case 298/87 *Smanor* [1988] ECR 4489.
30. Case 286/86 *Deserbais* [1988] ECR 4907.
31. OJ 1979 L33/1.
32. [1991] ECR I-2971.
33. Case C-85/94 *Piageme v Peters* [1995] ECR I-2955.
34. OJ 1992 L113/8.
35. EC Treaty art. 59.
36. See Usher, J. (1994) *The Law of Money and Financial Services in the EC*, Oxford: Oxford University Press, Chapters 4 and 5.
37. Case C-484/93 *Scensson and Gustavsson v Ministre du Logement* [1995] ECR I-3955.

Chapter 15

The Geopolitics of European Frontiers

MICHEL FOUCHER

Introduction

Border issues are back on the political agenda in Europe and for a geographer this is a very challenging matter. It is a fact of life that many internal borders have been upgraded or, in some cases, downgraded, into external political frontiers. Central and Eastern European countries are now framed by approximately 8000 miles of new political lines. In this respect, the so-called old continent is the newest of all, with more than 60 per cent of its present borders drawn during the twentieth century. Geopolitical instability is connected to people's perceptions of security and identity and, in this regard, several political borders are still problematic.

Several names for the same geopolitical object

The concept of border here refers to every category of border which has existed in Europe. Going beyond the rather large and imprecise notion of a limit, several types of territorial discontinuities should be envisaged.

In French, four different words (*frontière, front, marche* and *limes*) are widely used to describe frontiers. The etymology of the common word 'frontière' derives from 'front', a noun, and 'frontier', an old adjective no longer in use. 'Front' means front line in a military sense. Front lines are a common configuration in contemporary Europe, from Moldova to former Yugoslavia. In both cases, in the wars embarked on in a calculated way by one-party-State politicians, one of the main war aims was to enforce a redistribution of population according to newly imposed international borders. Another objective for these politicians was to gain political recognition from other competing centres of power, pursuing pure Realpolitik. Western troops have been assigned in ex-Yugoslavia, with IFOR (Implementation Force) and, since January 1997, SFOR (Stabilization Force) to act as border and front-line peace-keepers, along the Dayton-agreed demarcation lines, in a manner reminiscent of Cyprus.

The old concept of 'march' has returned in the context of the so-called security vacuum in Central Europe, when the status of neutrality is under growing criticism in some Western countries and in some political parties, and when European Union and NATO enlargement is seen, rightly or wrongly, as the necessary next step in the geopolitical reorganization of the continent, placing in a situation of jeopardy countries not invited to join the club. From the Baltic Sea to the Black Sea and perhaps part of the shores of the Adriatic Sea, a kind of 'Middle Europe' (*Europe médiane, Mitteleuropa*), an in-between Europe, is reviving, whose fate will be decided partly from outside the region, in Washington, Moscow, Bonn/Berlin and, perhaps, in London and Paris.

The precise meaning of the word Ukraine is 'march' or 'border area'; it will be very interesting to see how far a nation-state can be built in this border situation, with old pieces of the Russian, Austrian and Polish empires, 'unified' by Stalin half a century ago. The same question has been hesitantly posed in the area located around Minsk, Vitebsk and Brest-Litovsk; the lack of a specific Belorussian national identity is leading to a desire of reunification with Russia which, in economic terms, is too expensive. The process of nation-building is still under way in these areas which used to be a military glacis. Escaping buffer-zone status is an crucial ambition in Eastern Europe: it is not yet clear whether this is possible.

Another old concept has returned – the '*limes*'. The definition of this specific kind of border is problematic. The Roman *limes* was not a continuous closed wall. It was essentially a strategy aiming both at containing unwelcome migrants and at organizing trade with Romanized peoples and to bring them into a sustained peaceful relationship with the Empire. To the extent that a comparison is possible, the interaction between Europe as an organized union and its southern flank, from Morocco to Turkey, is similar to a '*limes*-style' strategy, where the border is closed to migrations and citizenship but open to trade, ideas and languages, in an asymmetrical relationship which remains a permanent source of tension.

Europe is becoming the site of a very complex system of specialized and selective borders which do not have the same geographical location and are situated, according to the objective pursued – economic integration, migratory protection, external security. Is the concept of 'frontierization' adequate to understand better what the essence of the European Union is?

A European Union 'without borders'?

A free-trade area changing into a genuinely single market (twelve members since 1993 and then fifteen in 1995); a Schengen area wherein people circulate freely, with seven states in 1996; a monetary union with, no doubt, eight members around the year 2000; projects of a common defence policy, with five or six countries involved: these are a few of the specific ways in which a Western Europe is taking shape, a reality intended by its founders to be free from internal borders. Existing borders must be made tension-free through the mutual understanding and economic and political interdependence; frontiers must be downgraded in association with a strategy of cross-border co-operation and by encouraging the movement of people, cultural goods and assets.

The promoters of the European idea wished to abolish national boundaries as barriers (here understood as tax points and police controls) by putting into effect the four freedoms contained in the Treaty of Rome and reaffirmed in the Single European Act of 1986. The free movement of goods, services, capital and people illustrates clearly the spirit animating the 'Founding Fathers'. They consciously aspired to put an end to the quarrels over where the frontier posts were placed, quarrels that had too long been allowed to set European peoples at odds with each other; it also explains why the borderlands of nations have become laboratories for opening nations to the outside world and nurseries for 'cultivating' committed Europeans.

The Fourth Freedom (freedom of movement)

Extending the trading areas in which goods may freely flow and increasing the exercise of joint sovereignty – sharing what used to be attributes of sovereignty – necessarily involved, *nolens volens*, a profound change of certain basic frontier functions. Should this be viewed as a political revolution, or as a minor adaptation involving the transfer to the new 'external' frontiers of the European Union those functions long carried out at the state frontiers which have now become 'internal'?

The long delay in putting the Treaty provisions into effect shows both the complexity of the criteria and of border control practices, and the reluctance of the states to give up fundamental sovereign rights. Thus the Schengen Agreement, allowing the free circulation at the frontiers of member states, signed on 14 June 1985, began progressively to go into effect only on 26 March 1995, five years after the application agreement was signed.

It was much easier to organize the free movement of goods than the free movement of people; as though people had become a cause of insecurity within Europe, the control of people's movements had been elevated to the main factor at stake in considerations of political sovereignty. This elevation has parallels with the United States which has also introduced computerized systems for its ports of entry.

Concerning the fourth freedom, one political change is significant, at a time when transfrontier networks and flows are increasingly important. To those in political power, the control of territory now seems less important than the control of those who claim the right to move about within it. Does this mean that we must consider this as a change from territorial sovereignty to a more jurisdictional sovereignty? There is a technical response to the problem of controlling the increasing flows of people, provided by the establishment of a data bank, the Schengen Information System for reciprocal, multilateral and ·instantaneous data exchanges between police forces, based in Strasbourg. This is a symbolic location since the Kehl bridge, over the Rhine river, has been crossed without a border check for several years now.

A twofold mutation involving the internal frontiers

On the one hand, frontier functions are disintegrating in a spatial sense. On the other hand, in certain respects, the entire national territory is now being treated as an expanded frontier zone.

Over a period of centuries, European states have steadily attempted to consolidate the 'envelope' of national territories, either by modifying their borders and making them permanent through wars which mobilized the population behind the aim of establishing internal unity. Treaties formalized new winner–loser relationships, and governments sought a single line on which to concentrate all the functions of interest to the state.

In the case of France, Vauban and Colbert laid down the basic principles of maintaining control of a territory and fortifying it against adversaries and competitors. Nation and territory, currency and market, were the end products of this typical political 'homogenizing' process, with the same spatial areas. French customs operations were nationalized in 1791, and gradually evolved into the 'immense administrative body' which, just before the Rome Treaty went into effect, was collecting 22 per cent of the tax revenue of the French state, and was still collecting 17 per cent in 1993 (petroleum, gas taxes and VAT on products from outside the European Community). The first modern German unification came about through a customs union, the *Zollverein*, just as it was through setting up the monetary union that the German Federal Republic was founded in 1949, from three occupation zones in which the Deutschmark became the single currency (extended in 1990 to the former East Germany).

'Organized' Europe is today a regional dimension of an inescapable process of globalization. The European integration process ensures that economic interdependencies are recognized and accepted, and these in turn provide a guarantee for peaceful relations. Economic flows in the EU in the middle of the present decade exceed in value 1400 billion ECUs. The customs service, as a tax service and economic agent, is no longer charged with managing 'legal routes' through which goods are allowed to pass. The customs function has been adapted to an economy with 'just in time' flows and zero stocks, and is therefore active in every production centre. At one time, customs was an operation entrenched at a national boundary; the customs function has now been prised away from the border and transferred elsewhere, wherever something useful is produced. Credit cards, the Swift network for movements of capital, purchase orders sent by fax and telex, all make for rapid financial business transactions. The business of financial management and the business of inspecting it are concentrated where the activities are to be found. In addition, the 'immaterial' service economy seems to be breaking out of spatial constraints. On the new French 'customs map', the greater part of the customs personnel is concentrated at the external frontiers of the Union, Switzerland, at maritime ports and airports. The number of customs offices in metropolitan France fell from 375 to 274 at the end of the year 1992. The customs officer has become an agent for regulating the national economy (checking to make sure standard requirements are met, controlling commerce with countries outside the European Union, seizing counterfeit goods). Customs work has gone inland to meet the goal of economic security.

The difficulty of managing the flow of people is shown with the help of a few simple, suggestive figures, frequently either unknown or ignored. Border crossings at the frontiers of the signatories of the Schengen Agreement total around 1700 million a year, when all three modes of travel, by land, sea and air, are combined. Movements across the borders of Germany account for 864 million; the figure is over 291 million for France. Spain, the Netherlands and Italy show figures of the order of 120–135 million. From the total of these figures, it has been estimated that 1200 million people annually cross the internal borders within the Schengen area.

Since the Treaty of Rome was signed, the number of border crossings is believed to have increased twentyfold. In the case of France, three-quarters of these crossings – around 230 million – are made by land. The administrative service involves the air and border police with some 50 stations and 762 accessible roadways and paths, taking into account the entire border from Dunkirk to Menton and from Le Perthus to Hendaye. Of this total number of 'passengers', made up for the most part of French citizens and citizens of neighbouring countries and other EU countries, only 5 per cent were travelling with a visa. A border zone 20 kilometres wide along the French borders has been set up, within which random mobile controls are carried out, as a compensatory measure to free movement. Control is becoming specialized: it is no longer the surveillance of a narrow linear system, but an extended system along the 'green frontier' – a term which in the language of specialized civil servants designates intervals between the main border crossings – and a reassignment of police forces to repression of illegal trafficking, directed at the illegitimate traffickers at their point of departure and the users at their destination. This means adapting surveillance to spatially extended networks at all the key points of a national territory and concentrating on the junction points of transportation networks.

The implementation of the Schengen agreements has introduced new spatial differentiations within the EU. Indeed, four types of situation have been identified. Seven of the nine signatory states have begun to implement the agreements: Germany, France, the Benelux countries, Spain and Portugal. Two other signatory states have postponed implementing them: Italy and Greece. Denmark, Sweden, Finland and Austria have signed, but have delayed implementation. Two states – the United Kingdom and Ireland – have not signed the agreements.

We therefore find within the European Union a Schengen 'sub-set' characterized by the following: freedom of movement for the regular residents of the Seven, of whatever nationality, including those from outside the Union; harmonizing visas of less than three months' duration (for a list of countries mutually agreed upon and using a single visa form issued by the consulates); common inspection at the external borders; liaison officers from each country on hand in the neighbouring countries to check on the application of the agreements concerning external borders. Beginning in March of 1995, entries at airports have been reorganized to separate the flow by country of origin (around 200 million crossings annually).

The United Kingdom seems determined to maintain the control of individuals who come from countries outside the Union, and this forces Ireland, which has had no frontier control with its neighbour since the 1920s, to comply with British practice. Finally, the two Scandinavian countries, which both have an open land frontier with a state which is not a member of the European Union – Norway – must harmonize the agreements on crossing this atypical external border. The basis for free movement varies depending on the degree to which the countries participate in these agreements. As is the case where currency or institutional reforms are concerned, the Europe of fifteen member nations contains various configurations, divided according to more or less temporary technical and functional frontiers. Differences in harmonization also relate to the problems of external frontiers, primarily because of the geographical diversity of migratory pressures.

Migratory borders

An attentive reading of recent European history shows the close correlation between the more remarkable political changes and migration data. Since 1989–90, the area involved in migration increased to cover the entire continent. It was partly a migratory pressure based on political and economic factors which brought down the old geopolitical order. The breach opened at the Austro-Hungarian frontier, following a bilateral German-Hungarian agreement for the benefit of East Germans, which was then widened by the dismantling of the Berlin Wall. The fourth freedom – free movement – was eagerly anticipated by the East Germans; one of the first new branches of service activity to blossom in East Berlin was travel agencies.

The Iron Curtain served as a very effective migratory barrier in the middle of the continent, with limited exceptions for East Germans and the states which allowed limited departure of their German minorities in exchange for cash (Romania and the USSR). One of the objectives of German *Ostpolitik* from the beginning was to help German minorities to leave (1 million returned between 1950 and 1969, 1.3 million between 1970 and 1989), on the strength of Article 116 of the German *Grundgesetz* which defines the criteria for allowing people to return to their homeland, either on the basis of their German citizenship before 8 May 1945, or because they are ethnic Germans.

The Hungarian government decided in May 1989 to dismantle a portion of the Iron Curtain between Sopron and Hegyeshalom. This was a result of a policy of German-Hungarian *rapprochement* which began as early as 1986, when the government of Janos Kadar 'became aware of the advantages of treating its Germans in an exemplary manner' as Timothy Garton Ash recalls it. Hence the granting of an unconditional credit of 1000 million DM in 1987 – a 'signal flashed to the other states of the Soviet bloc'.[1]

Indeed, Bonn was saying that the new *Ostpolitik* meant financial encouragement to its compatriots to stay put. But, paradoxically, it fostered both the negotiated dismantling of the Iron Curtain and a new migration of minorities to Germany, so that Central and Eastern Europe have never had so few German residents as they have today. This repatriation largely explains why Germany in 1992 received two-thirds of all the migrants entering Western Europe.

In addition to this specific national situation, the general European migration problem has changed. While the states of Western Europe have kept their frontiers closed to large migrations of labour for the past twenty years, the movements have nevertheless continued. This is for various reasons – the serious political and economic situation in the country sending out its workers, the ineffective nature of the restrictive measures in the country of destination, the admission of family members joining those already settled and an increasing number of clandestine entries.

Some countries, formerly the source of migrants in Southern Europe, have become destination countries: Spain for Moroccans, Italy for Moroccans and Albanians and Greece for Albanians. Countries such as Poland have seen the arrival of people from countries further east in Europe and their transit to Germany. The European Union numbers around 20 million foreign residents, 6 million of which are from other European states – 4–5 million having come from Turkey and the countries that formerly were part of Yugoslavia, 3–4 million being from North Africa and other parts of Africa and 1 million from Asia. The change in migratory patterns, marked by constant pressure from the peripheries of the Union – Turkey first and foremost, South Asia, through Russia, the Balkans, especially Albania and Romania – will probably lead to the establishment of a quota policy, such as that of Switzerland.

We may speculate about the extent to which the management of migratory flows originating outside the Union will become a central element defining the de facto frontiers of 'organized' Europe. German immigration policy, through a series of restrictive bilateral agreements on the right of asylum, provided a 'migratory space' the borders of which are not at the Oder-Neisse but further east and south-east, including states of Central Europe already associated with the Union and considered as likely candidates for admission. Since July 1993, asylum seekers arriving in Germany from Polish territory are sent back to Poland, and Poland is provided with financial compensation for the costs incurred by reinforced control at the frontiers and by the reception of refugees, mostly Romanians.

Comparable agreements were signed in 1994 with the Czech Republic, Hungary and Romania, and have been negotiated with Bulgaria, to reduce the number of Turks arriving. The eastern frontiers of the states neighbouring Germany and the states nearer to the sources of emigrants are, if the agreements can be applied, on the way to operating like the immigration frontiers of the European Union, way beyond its present frontiers. This is an unstable response to the contradiction between greater openness of internal frontiers and the reinforcement of controls at the external borders, at a time when these neighbouring states are being brought into association with the Union in a transitional process leading eventually to full EU membership.

The external frontiers of the European Union today, from the Strait of Gibraltar to the Kola Peninsula

Travelling from Tangier to Murmansk by way of Istanbul and St Petersburg, one discovers the wide diversity of the immediate neighbours of the European Union. The only element shared equally by such different states (and also regions and cities) is the interaction that makes movement towards the 'poles' – represented by the cities and states of the Union – so attractive. The various neighbourhoods of the EU can be classified in seven distinct groups, from the south-west to the north-east.

(1) To the south-west, the Maghreb, a western Arab-Berber island between the Mediterranean and the Sahara, has an average standard of living between five to ten times below that of the neighbouring states in the Iberian peninsula and of Italy. Their aggregate gross national products are 4 per cent of that of the fifteen members of the EU. The Union is both their biggest customer and their largest supplier, so that their de facto integration in the European market is further advanced than the states of Central and Eastern Europe.

The Union is the main provider of loans to North Africa and, notwithstanding a widespread idea to the contrary, 'the opening out to the east' has not resulted in a decrease in the flows of financial assistance from Brussels. Direct financial aid granted for the period from 1991 to 1996 increased threefold over that from the previous agreement: 1425 billion ECUs from the Community budget and 1400 million ECUs for the loans of the European Investment Bank (EIB). This was a deliberate decision aimed at establishing a viable EU Mediterranean policy, putting into effect the conclusions of the Matutes report. The Barcelona Conference in November 1995 confirmed a grant of financial assistance amounting to 5.5 billion ECUs (from 1995 to 1999). This figure is comparable with the amount granted to the countries of Central and Eastern Europe together: 7.1 billion ECUs.

Economic strategies of delocalization seem to be a response to two questions: how to slow down migrations? how to foster growth in the countries ready to export workers? The Maghreb countries stand to benefit from the answers. But competition from eastern Europe comes into play because the cost of labour there is just as low. North Africa has long been an area of strong French cultural and economic influence, but it is now also the scene of increased activity on the part of the Italians and Spaniards who have been investing heavily in Morocco since 1991. A new geo-strategic association bringing together the nine states bordering on the western Mediterranean is thus taking shape, including five states which belong to the Maghreb Arab Union, but at present is held back by the civil war in Algeria.

This group of EU neighbours offers at least two special features. First, the countries on the southern shore of the Mediterranean have no prospects of membership. The most they can hope for is a long-term 'partnership', with the end result of inclusion in a free trade zone some time around the year 2010. Second, the southern neighbours of Europe are not the most important economic partners of the southern European states; they account for only around 28 per cent of the total exchanges, excluding petroleum. The poor transport links across the Mediterranean and the greater attraction of the north-western part of Europe help explain this.

The Maghreb is an integral part of the European geo-economic space, but it is prevented from sending its workers to Europe, as it did in the past. The migratory blockage is a significant factor in the social crisis in Algeria, which has resulted in a civil war, and one of the stakes involved in this war is the nature of the country's relation to Europe. Two other states remain the object of attention, partly because of their geo-strategic position: Morocco to the south of the Strait of Gibraltar and Tunisia to the south-west of the Sicilian Channel.

European strategy seems hesitant, but the question is posed about whether it is in the interests of the EU to see bonds develop horizontally between the states on the southern Mediterranean shore as the first steps towards forming an anti-Western political coalition? The obvious interest of southern Europe and the European Union is to help the Maghreb countries to become the most modern region of the Arab-Muslim world.

(2) Italy, by virtue of its Adriatic coast frontier, its land border with Slovenia and its complicated relations with Slovenia and Croatia (involving the status of an Italian minority in Istria, the question of indemnities for the Italians in Slovenia and, until recently, the Italian veto of an agreement to associate Slovenia and the Union), is directly involved in Balkan and Central European problems. Albanian migratory pressure is one expression of this. The Strait of Otranto spans one of the greatest gulfs in living standards in Europe, the impact of which is intensified by the images broadcast by commercial Italian television, showing Albanians the good life in Italy. This strait is patrolled by an Italian air and naval group tracking the fast boats crossing at night with illegal immigrants. The Italian state authorities, influenced by the American example of management of the coasts of Florida, are considering sending logistic and police brigades and humanitarian aid to Albania and maintaining surveillance of the Albanian ports. In Rome, it is thought that policing the Otranto Strait might be one of the future missions of the *Euromarfor*, an Italian initiative, bringing together France, Spain and Italy in an effort to promote security in the western Mediterranean area.

Italy seeks to develop positive relations with new neighbour states, such as Slovenia which became independent on 7 October 1991, and which finds sympathies in Friuli-Venezia Giulia. The Italian government has been active diplomatically in south-central Europe, with its *Pentagonale* initiative in 1989 (which was oddly broadened to include Poland in 1991 – and thus transformed into a *Hexagonale*). It serves mainly as the framework for co-operation and for the promotion of Italian commercial interests in the region south of the Danube. Italy ranks second behind Germany of the countries of the Union in export market share (around 15 per cent) in this area.

(3) In the south-eastern corner of the European continent, the Union is represented by a single isolated country, Greece, whose immediate neigh-bours present a particularly complex and volatile picture: Albania,

Macedonia, Bulgaria, land and sea frontiers with Turkey and, beyond them, the Cyprus question. The Union seemed to have here in Greece a valuable anchor. In fact, Greek governments have, for internal political reasons, sometimes seemed to adopt policies against their economic interests, and failed to extend Greek influence to its immediate neighbours. Greece, which derives over 6 per cent of its GNP from the European Union budget, is no doubt a poor relation in the Union, but it is perceived by its neighbours as a prosperous and attractive country. The migratory pressure on the north-west Greek border from Albania is strong. In Macedonia, people deplore the commercial blockade which blocks the Vardar corridor, although Macedonians are still orientated towards Salonika, rather than Belgrade. Relations with Bulgaria are better, and the valley of the Struma is busy with Greek-Bulgarian trade.

Greek-Turkish relations are, as ever, fraught with difficulties. The 1923 Lausanne system is unstable: the Cyprus question, the problem of sea and air frontiers, the fate of the Muslim minority in Thrace, the rivalry between Athens, Salonika and Istanbul, the great metropolis of south-eastern Europe (whose market zone today extends as far as Bucharest, Constantza, Odessa, Kiev – and even further north in the direction of Russia and Lithuania). The signing of a customs union between Brussels and Ankara has been made possible in exchange for a Community commitment to conduct negotiations between the Union and Cyprus immediately after the 1996 Intergovernmental Conference.

It is more than probable that outside powers will continue to have decisive influence over the destinies of south-east European countries. The United States seek to extend their zone of military influence (NATO) northward to include Albania, Macedonia (American soldiers under UN authority have been at the Serbian frontier of Kumanovo) and doubtless, some day, Romania. This presence is opposed by the Russians and Serbs seeking to maintain anti-Turkish alliances in the Balkans and to set up a political Morava-Vardar axis (Belgrade/Athens and Sofia/Bucharest) against 'horizontal' projects (infrastructure projects among others), from Tirana to Istanbul by way of Skopje and Sofia. Hints of the 'great game' of the Balkans at the end of the nineteenth century can still be seen.

In the larger scheme of European integration, south-east Europe is approaching an advanced state of Balkanization, and is continuing to fall behind. Finally, the Greek government feels a peculiar responsibility for some 600,000 Greek 'Pondi' – Greeks in the Black Sea region in southern Russia – and plans for their return to the mother country are under way.

As an exclave of the European Union, Greece – where application of the Brussels Directives is slow – has recently taken on added geopolitical importance with the political changes in Albania and Bulgaria, as well as with the partitioning of the former Yugoslavia and the setting up of successor states. Greece is the only state in the Union which might be directly involved in a territorial dispute if – supposing that the independence of Macedonia is confirmed – the government in Athens felt obliged to take a stand on its traditional position, refusing to recognize a state bearing the Hellenic name of Macedonia.

Greece has another geopolitical characteristic that sometimes goes unnoticed but which might prove of strategic importance in a pan-European Union policy: it is the only Orthodox state in the Union. If we consider it pertinent to differentiate between the continent's historic religious units, it is clear that, some time in the future, the recomposition of Europe will imply a dialogue between two very distinct types of society, a dialogue that until now has proven difficult. Here, Greece may find its role as a 'bridge', analogous to that of Spain which, in 1992, demonstrated its usefulness as an interface between the European Economic Community and Latin America. The priority for Brussels is to strengthen the Greek 'bastion' in an unstable region.

(4) The admission of Austria into the Union is a consequence of the geopolitical upheaval of 1989. The status of neutrality imposed by the 1955 state treaty made Austria a buffer-zone between two blocs. A review of this geo-strategic in-between situation began with the 1995 admission to the EU. The Union's Austrian borders touch the southern part of the Czech Republic, the western part of Slovakia (its capital Bratislava only 40 km away), the western part of Hungary (with Budapest 250 km from the border), and the northern part of Slovenia. Vienna is reasserting a former role as an economic and urban pole of the front rank, backed by a strengthened Austrian economic commitment to Central Europe. In these circumstances, the eastern part of Austria experiences migratory pressures, and the eastern frontiers are both more open than formerly and more closely watched. Multinational projects centred on the Austro-Slovako-Hungarian tripoint remain in the planning stage until the Hungarian-Slovak bilateral issues have been resolved. The western part of Hungary – the Györ region – has been benefiting the most from economic openness, along the new Vienna-Budapest motorway. This is a virtual cross-border zone, in which Vienna, the former 'gate to the East', is strengthening its role as a communications and management centre (with 800 main offices of Western and Asian firms and 400 foreign press correspondents covering Central Europe). This assertion of a central role implies conflict with the financial ambitions of the Budapest market. Austria is also insisting that its neighbours bring their nuclear power plants (Krsko in Slovenia, Bohunice and Mochovce in Slovakia, Temelin in the Czech Republic) up to Western standards; and *Ostökofonds* – the Austrian fund for ecology – is directed at limiting pollution originating from neighbouring countries.

(5) To the east and south-east of the reunified Germany, Central Europe proper contains four states: Poland, the Czech Republic, Slovakia and Hungary. Seen from Berlin, only 80 km from the Oder-Neisse frontier, Central Europe is an important export market but one which is not completely dominated by German products (40 per cent of the market share of OECD countries on average). It is an area for industrial sub-contracting with qualified workers; and it is made up of a group of countries to be brought, for security reasons, into the European Union as soon as possible, at a cost to be negotiated and shared with the other member states. Germany aspires by the turn of the century to cease providing the eastern frontier of the Union and become, like France, a country surrounded by Member States. The security horizon is also perceived as extending as far as the Bug – on the eastern border of Poland.

Since the time of Adenauer, the Bonn government understood that recognition of the Oder-Neisse line was necessary to gain the support of the Western allies for reunification of the Germany between the Rhine and the Oder. The reluctance of Chancellor Helmut Kohl to commit himself publicly on this subject in 1991 was less the expression of a reservation in principle as an attempt to mould domestic public opinion:

> The strategic plan had been to show the nationalist right and the last die-hards, among those expelled, that this was simply the price that Germany had to pay for unifying the Germans between the Oder and the Rhine. He could then say, in effect, to the authentic revisionist nationalists: is it really the sabotage of German re-unification that you are asking for?[2]

Final ratification of the definitive treaty on the borders with Poland in the autumn of 1991 implied a new bilateral German-Polish treaty, and protection of the rights of minorities in Upper Silesia.

The Oder-Neisse frontier represents a gulf of the same magnitude, in terms of living standards, as the boundary separating southern Europe from the Maghreb. None of the states in either region can hope for financial help on the scale (110 billion DM a year) provided for the former German Democratic Republic. But, of all of the states in the former Eastern bloc, they are the first in line, both physically and figuratively speaking, to benefit from a positive frontier effect. Bonn and the German Länder are actively encouraging the formation of Euro-regions from the western upper reaches of the Neisse valley to southern Bohemia (Nisa, Elbe/Labe, Egrensis, Sumava).

Opposition by the Polish and Czech governments to these initiatives, which involve recasting administrative structures in their centralized countries, is apparent. The Prague and Warsaw governments do not regard cross-border regions as instruments for integration with the Union, nor as excluding other forms of local transfrontier co-operation. But financial help from the European Community for infrastructures and for improving the environment is welcome. PHARE and the specific programme CROCO – Cross-Border Co-operation – amount to 150 million ECUs annually.

(6) Further north, the Baltic Sea, once divided by a maritime section of the Iron Curtain, now serves as a framework for close co-operation between, on the one hand, Denmark, Germany, Sweden and Finland and, on the other, Poland, the Baltic states and north-western Russia. The increasing number of relations between Helsinki and Tallinn, Stockholm and Riga, makes the Baltic states the first beneficiaries of support for reorienting trade towards the West, and of a progressive emancipation from Russian tutelage. Here again, the frontier effect is at work, particularly illustrated by the migration of workers back and forth between Estonia and southern Finland. The membership of two Nordic states does not prefigure the formation of a specific community of states bordering the Baltic Sea, because the difficulty that the three Baltic states have defining common approaches complicates efforts to organize co-operation. Since the Ronneby conference, close links have been established between Finland, Estonia and the *oblast* of Saint Petersburg and Karelia in efforts to protect the environment. In May 1996, Sweden organized the Visby conference, on the Baltic island of Gotland, in which nine countries, including Russia, participated with the objective of securing more effective financial support from the European Commission. Some border issues, such as those between Russia and Estonia, have been addressed, and the question of the integration of the three Baltic republics. But the hidden agenda was mostly about the new geo-strategic setting of the Russian neighbourhood of the Baltic rim.

(7) One notable development since 1995 is that the European Union now has a nearly 1313 km long direct frontier with Russia. Finland is interested in PHARE funding for improving transit facilities (particularly towards the zones in Russian Karelia where forests are harvested) and for protecting the environment,[3] and especially to control pollution from nuclear plants in the Kola peninsula. To the east of the Vyborg isthmus, the municipality of St Petersburg is, as in the past, promoting itself as the 'Russian Gate to Europe' in competition with Riga and Moscow. But the strategic future of this new external border of the Union is the main issue.

By way of conclusion

The many forms of interaction at the current frontiers of the European Union are present in virtually all the major geographic locations within the Greater European area, understood as the geo-economic zone including the western and central Mediterranean, south-eastern Europe (including Turkey), Central Europe and Russia.

The new phenomenon in the contemporary scene is the emergence of the functional frontiers, geographically dissociated from the official frontiers of the EU. We have seen this is true for migrations – Central Europe once served as a transit zone but is now a barrier zone filtering out clandestine immigrants. This is also true for institutional membership since the high costs of enlarging the Union are going to encourage more realistic approaches than some public declarations of a future Europe open to all. Finally, while this is not the focus of this concluding chapter, other frontiers are involved in extending NATO membership – new security frontiers. Paradoxically, NATO membership is proposed for states which have the fewest objective security problems, namely those of Central Europe.

As in the past, the security of these states is considered in terms of external preoccupations, especially the security of Germany and the maintenance of the interests and influence of the United States. Moscow clearly prefers the status quo, with what is considered by some as a 'security vacuum', but perceived by the Russians as an external military glacis – or, in other words, a buffer-zone much like that of Finland or Austria before 1989. In a continent whose present, and recent, frontiers are still likely to be relocated or be the subject of conflicts, the search for security may for the time being take the form of diplomatic stabilizing action. This requires recognition of all existing frontiers, and extending good neighbour and local co-operation agreements. Around one hundred such agreements have been signed since 1990, confirmed by the Stability Pact adopted on 20 March 1995 at the second Paris Conference. Franco-German co-operation serves as a model for peaceful frontier interaction. This model is used in other cases of bilateral relations, with a turbulent history, in efforts to achieve a Europe free from militarized front lines. Tragic borderline quarrels requiring a negotiated settlement are present and will remain on the agenda for a long time in south-eastern Europe and on the borderlands of the Russian Federation.

A final remark is not about space but time. If Europe is destined to be a very complex system of interaction, time is an element of such a complexity. Borders are time inscribed into space or, more appropriately, time written in territories. Since 1989, European peoples have been asked to live and act at the same 'meridian time', although it is not clear if the new meridian is drawn through Brussels or . . . Washington. But there is still the specific political time of the nation-state building process, which is sometimes drawn on the ground with blood. In this political time, border is an idol asking for human sacrifices, in the words of the Triestine author, Claudio Magris. There is also the world time of the global economy which is putting pressure on national borders to dissolve them into free trade areas and areas with uniform consumption patterns: the curious notion of 'market democracy' should, according to this vision, be the new paradigm. There is now an attempt to adhere to a more specific European time, where the old state borders could be overcome by negotiation and through democratic procedures. The historical and geopolitical diversity of Europe explains the difficulty of creating common values and procedures. The emphasis on hardening external borders of the Union is related to an ambitious collective desire of the Europeans to remain the 'master of the clock' as they used to be in the past when the world was symbolically structured around the Greenwich meridian.

Notes

1. Ash (1995), p. 281.
2. See *ibid.*, p. 267.
3. See the map indicating sources of pollution in Karelia, in Foucher (1993), p. 248.

Bibliography

Ash, T. G. (1995) *Au nom de l'Europe, L'Allemagne dans un continent divisé*, Paris: Gallimard.

Foucher, M. (1991) *Fronts et Frontières, un tour du monde géopolitique* (second edition), Paris: Fayard.

Foucher, M. (ed.) (1993) *Fragments d'Europe. Atlas de l'Europe médiane et orientale* (second edition), Paris: Fayard.

Foucher, M. (ed.) (1996), *Les défis de sécurité en Europe médiane*, Paris: FED/ Documentation Française.

Foucher, M. and Oyarzabal, I. (eds) (1996), *Visions of Europe*, Madrid: Foundation BBV.

Index
